An Atlas of World Affairs

The economic, social and environmental systems of the world remain in turmoil. Recent years have seen possibly irrevocable change in the politics of Europe, Asia, Africa and Latin America.

Entirely revised and updated, the eleventh edition of *An Atlas of World Affairs* describes the people, factions and events that have shaped the modern world from the Second World War to the present day. International issues and conflicts are placed in their geographical contexts through the integration of nearly one hundred maps. The political context provided for current events will be invaluable to all those uncertain about the changing map of Europe and Africa, conflicts in the Middle East, and the appearances in the headlines and on our television screens of al-Qaeda, Chechnya, the Taliban, Mercosur, Somaliland, Kosovo, AIDS, OPEC and Schengenland. Critical new issues are covered, including the war on terrorism, nuclear proliferation, European Union expansion, and the pressing environmental concerns faced by many sovereign states. This edition provides guidance through all these recent changes (and many more).

This book offers up-to-date coverage of all regions in great detail. It contains an objective and concise explanation of current events, combining maps with their geopolitical background. It provides a clear context for events in the news, covering the Middle East, Korea, China, the European Union, east Africa and every other part of the world. Revised and in print since 1957, *An Atlas of World Affairs* continues to provide a valuable guide for the student, teacher, journalist and all those interested in current affairs and postwar political history.

Andrew Boyd began his acquaintance with international affairs in 1946, when as a British liaison officer he attended the very first sessions of the United Nations (his other books include three about the UN). He travelled widely and reported on international affairs while writing on world affairs for *The Economist* for 37 years.

Joshua Comenetz has used cartographic methods to visualize spatial data and explain the causes and effects of international conflicts, demographic change and natural disasters since 1990. As a consultant he has solved problems in areas ranging from political redistricting to ethnic and religious mapping, and he has taught international relations and geography at university level.

An Atlas of World Affairs

Eleventh edition

**Andrew Boyd and
Joshua Comenetz**

Routledge
Taylor & Francis Group

LONDON AND NEW YORK

First published 1957 by Methuen & Co. Ltd
Second edition 1959
Third edition 1960
Fourth edition 1962
First published by Methuen as a University Paperback (fifth edition) 1964
Sixth edition 1970
Seventh edition 1983
Reprinted 1985
First published by Routledge (eighth edition) 1987
Reprinted 1989, 1990
Ninth edition 1991
Tenth edition 1998
Eleventh edition published 2007 by Routledge
2 Park Square, Milton Park, Abingdon, Oxon OX14 4RN

Simultaneously published in the USA and Canada
by Routledge
270 Madison Ave, New York, NY 10016

Routledge is an imprint of the Taylor & Francis Group, an informa business

© 2007 Andrew Boyd and Joshua Comenetz

Typeset in Amasis and Akzidenz Grotesk by
RefineCatch Limited, Bungay, Suffolk
Printed and bound in Great Britain by
The Cromwell Press, Trowbridge, Wiltshire

British Library Cataloguing in Publication Data
A catalogue record for this book is available from the British Library

Library of Congress Cataloging in Publication Data
A catalog record for this book has been requested

ISBN10: 0–415–39168–7 (hbk)
ISBN10: 0–415–39169–5 (pbk)
ISBN10: 0–203–96752–6 (ebk)

ISBN-13: 978–0–415–39168–9 (hbk)
ISBN-13: 978–0–415–39169–6 (pbk)
ISBN-13: 978–0–203–96752–2 (ebk)

To Andrew

Andrew Boyd first brought *An Atlas of World Affairs* to our shelves in 1957. Through ten revised editions, it has become canonical for students, teachers, journalists and anyone with an interest in postwar politics and current affairs. His comments on the ever-changing political, economic, social and environmental systems of the world can be accused of nothing but honesty, without a hint of bias.

In between writing *An Atlas of World Affairs* and many other works of a similar genre, he excelled as a journalist for *The Economist* and was a fantastic father and grandfather. It was only true to his nature that he bowed out after ten editions, claiming in sublime modesty that he was not well enough to continue with another edition. Andrew Boyd sadly passed away in January 2003, but will forever be remembered on both a personal and professional basis as a talented and extraordinary gentleman.

Written by Claire

Contents

viii   *Contents*

Foreword

The world changed dramatically in the past decade. The 'war on terror' replaced interstate war and focused attention on the Middle East. Nuclear proliferation brought Iran, North Korea, Pakistan and India into the headlines. The US invaded Afghanistan and Iraq, war engulfed central and eastern Africa from Congo to Sudan, and Somalia collapsed. East Timor broke away from Indonesia, as did Montenegro from Serbia (and Kosovo sought to do the same). China absorbed Hong Kong and rose as high as second place in the world economic ranking; India aimed to follow. The European Union grew to include 27 nations, and NATO expanded into eastern Europe, both incorporating the former Baltic republics of the Soviet Union. This revised edition of *An Atlas of World Affairs* provides guidance through all these recent changes (and many others).

An Atlas of World Affairs was first published in 1957. The first edition's foreword included this passage:

Anyone who tries to set down some of the complexities of this changing world in simple form is indebted to the pioneer work of Mr J. F. Horrabin. The admirable simplicity of his pre-war *Atlas of Current Affairs* is hard to emulate nowadays . . . But, like Mr Horrabin's, this is still 'an exercise in the art of leaving out'.

Fifty years on, in the face of ever-increasing complexity, the aim is still to select what is relevant, and to explain a changing world's problems with the aid of simple maps and brief notes.

Notes

For the sake of brevity, the United Kingdom of Great Britain and Northern Ireland is usually called Britain; the Netherlands is called Holland, and so on. The United States of America may be America, the US or the USA. The former Union of Soviet Socialist Republics may be the Soviet Union or the USSR. The European Union, formerly the European Community, appears as the EU or the EC.

Distances are expressed in miles. One mile is roughly 1.61 kilometres. One nautical mile is roughly 1.85 kilometres. To convert square miles into square kilometres, multiply by 2.59. The ton and the metric tonne are roughly equivalent. There are about seven barrels of oil in a ton.

An italic number in brackets – e.g. *(68)* – is a cross-reference. The number refers to a section and its accompanying map or maps, not to a page. So do the entries in the index.

1 People and Pressure

The human race has trebled its numbers in less than one lifetime. In 1930 there were about 2 billion people. Now there are about 6.5 billion. The annual increase is reckoned to be about 75 million.

Today the two most populous countries, China and India, have between them more inhabitants than the whole world had in 1930. Both countries are in Asia, which, for countless centuries, has always contained more than half of humanity. Together, Asia, Africa and Latin America now contain almost five-sixths; and that proportion is still rising, because many countries in those regions have relatively high rates of population growth. By contrast, rates of natural growth are generally low in European and East Asian countries, and in some of them growth has stopped, although immigration may keep the population figures rising.

Growing at 2% a year, a population doubles in 35 years. A few decades ago most countries outside Europe and North America exceeded this rate. In recent years growth rates have dropped around the world, and now few countries outside Africa are growing at such a pace. The majority of the world's governments have adopted policies aimed at limiting population growth, but the results have varied. The limiting of family size has been actively discouraged by some religious authorities and some national governments, though even theocratic governments such as Iran's have moved to reduce population growth. As birth rates decline, migration and ageing will continue to increase in importance as the world's primary drivers of demographic change.

Historically, population growth was sometimes seen as a sign of national strength. Forests were meant to be felled and plains to be tilled. This was especially true in the Americas, vast and thinly populated in the sixteenth century – the more so after genocide and diseases brought by European explorers and settlers decimated the indigenous population. When Thomas Malthus warned in 1798 that population growth would eventually overwhelm the earth's capacity to provide food, he was observing what was then the zone of fastest natural increase in the world: the United States. Intensive agriculture has ensured that his prediction has not come true, nor is there any immediate prospect of a worldwide food shortage. But more people means more pressure on natural resources. In the past few decades, some effects of rapid population growth have become more visible and more alarming: soil erosion, overgrazing, destruction of forests, enlargement of deserts; in the oceans, the devastation of fish stocks; in cities, the multiplication of crowded slums.

The last of these is associated with a prominent feature of our time: the growth of giant cities. In 1950 there were only six cities or conurbations that contained as many as 5 million inhabitants. Now half of the world's population lives in urban areas and there are at least twenty whose population exceeds 10 million. Some of these include huge slums packed with people who have been forced to leave overpopulated rural areas.

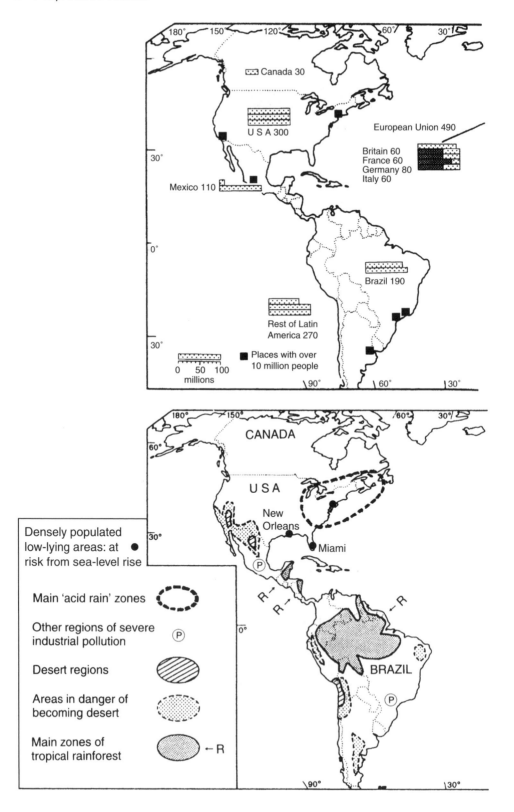

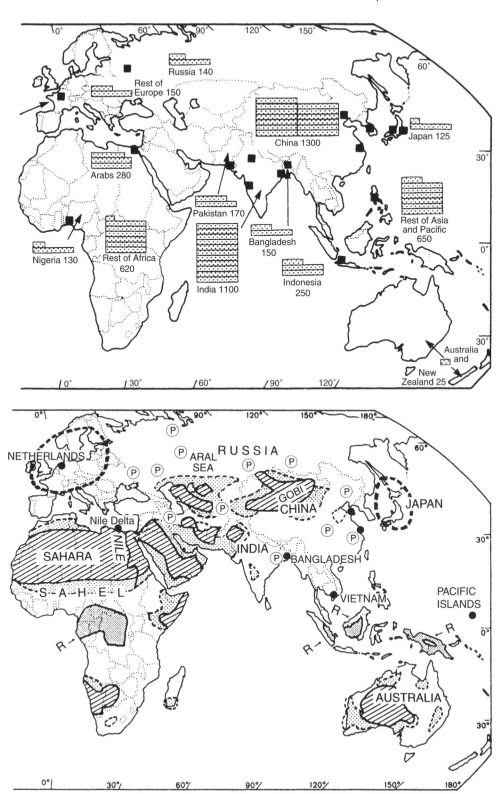

Russia 140

Rest of
Europe 150

Arabs 280

Nigeria 130   Rest of Africa
620

Pakistan 170

India 1100

China 1300

Japan 125

Rest of Asia
and Pacific
650

Bangladesh
150

Indonesia
250

Australia
and
New
Zealand 25

NETHERLANDS

Ⓟ   Ⓟ   ARAL
SEA   RUSSIA   Ⓟ

Ⓟ   Ⓟ   Ⓟ

Ⓟ   Ⓟ   GOBI   Ⓟ
CHINA   JAPAN

Nile Delta   Ⓟ   Ⓟ

SAHARA   NILE   INDIA   Ⓟ   Ⓟ
Ⓟ BANGLADESH

S – A – H – E – L   VIETNAM   PACIFIC
ISLANDS

R→   R→

R→

AUSTRALIA

Ill-chosen policies have contributed to the loss of forests and the growth of deserts. In some countries, uncontrolled logging and misdirected subsidies have caused disastrous deforestation. The 'great leap forward' that China's Maoists ordered from 1958 to 1962 left vast tracts of land deforested and desertified, as well as causing 28 million deaths from famine. In the 1990s, desert areas around the Aral Sea were still expanding, and that sea itself was drying up, as a result of Soviet 'planning' *(20)*. High population growth in arid regions from the Middle East to the south-west US has increased the pressure on ground-water resources. In some dry areas, ancient deposits of underground 'fossil water' are not replenished by rainfall, and aquifers are being used up or 'mined'. The most notable example is Libya's Great Manmade River, a system of pipelines that supplies coastal cities and agriculture from Sahara aquifers.

The 'acid rain' that damages forests comes mainly from emissions of sulphur dioxide. These have been cut by a third or more since 1980 in western Europe and North America but are on the increase in industrializing Asia. Airborne pollution from the burning of coal or oil (fossil fuels) can also endanger health. And by the 1980s, there were fears that carbon dioxide emissions from fossil-fuel use would cause a 'global warming'. If the ice in Antarctica and Greenland *(77)* then melted, the sea level would rise and inundate densely populated coastal areas. The Kyoto Protocol of 1997 sought to reduce global warming through internationally agreed national limitation of carbon dioxide emissions. Little has been accomplished, partly because the Protocol was not ratified by the US, the world's largest economy and greenhouse-gas producer. On another kind of harmful emission, however, action was taken. Man-made chlorofluorocarbons (CFCs) had made a hole (over Antarctica) in the ozone layer in the stratosphere that shields the earth from ultra-violet rays. By 1996 production of CFCs in the OECD (Organization for Economic Cooperation and Development) countries was phased out, and the hole has stopped growing.

The future extent of a manmade 'global warming' is still debatable. The experts who predict it and measure its beginnings do not all agree about its speed, scale and effects, but the level of carbon dioxide and other atmospheric greenhouse gases has risen rapidly in recent decades. Sceptics point out that, in the past, temperatures have often changed without any help from mankind, so the changes now detected may not be wholly manmade. But those who seek to curb the use of fossil fuels have other arguments. Although many new oilfields may yet be found, these fuels are finite. The long-term aim must be to avert the exhaustion of finite fuels by making maximum use of renewable energy sources. An encouraging development is that the costs of harnessing wind and solar power have been falling.

2 Economic Groupings

There have been dramatic changes in the balance of economic power. When that power lay mainly in North America and Europe, and the only challengers were the Soviet Union and, later, Japan, the simple concept of a world consisting of a rich 'north' and a poor 'south' seemed valid. It is less valid now that some southern economies are growing much faster than northern ones, while the Soviet collapse glaringly revealed the relative poverty of eastern Europe and Russia.

A 'rich north' can still be identified. The members of the OECD (Organization for Economic Co-operation and Development) comprise 22 European states, along with Australia, Canada, Japan, Mexico, New Zealand, South Korea, Turkey and the United States. A few years ago, these nations, with only a sixth of the world's population, were rated as producing three-quarters of its total output. They loom less large now that such institutions as the World Bank use 'purchasing power parities' ('PPP') in comparing one nation's GDP (gross domestic product) with another's. The PPP measure uses theoretical exchange rates reflecting cost-of-living variations rather than market exchange rates. This increases the size of poorer countries' economies because their cost of living is generally well below that of wealthier countries. By this measure, the output of the OECD states is now about half of the world total.

The older method of basing comparisons on market exchange rates had disguised the rapid growth of some Asian economies. In recent years China's GDP was achieving an annual growth of around 10%; South Korea, Indonesia, Malaysia, Thailand and other 'newly industrialized countries' (NICs) usually had much higher growth rates than the OECD states. Between the 1950s and the 1980s, Asia's most spectacular performance had been that of Japan; its GDP became the world's second-largest *(58)*. Now, by the PPP measure, China has moved past Japan, and is expected to pass the US in several decades. Although China is far behind Japan in income per head by any measure, it has ten times Japan's population. The same process of rapid economic growth applied to a very large population will soon lift India into third place in the world.

Both large and small states have been discovering the advantages of regional and bilateral free-trade pacts. In western Europe, which took the lead, the original six-nation 'common market' has expanded to make free trade broadly effective across the continent. Nowhere else has there been such an ambitious move towards integration as the forming of the 25-member European Union, but the basic idea of regional free trade has been widely adopted. A free-trade area comparable in scale to the European one has appeared in North America, where the Canada–United States pact that took effect in 1989 was enlarged in 1994 by Mexico's inclusion in the North American Free Trade Agreement (NAFTA).

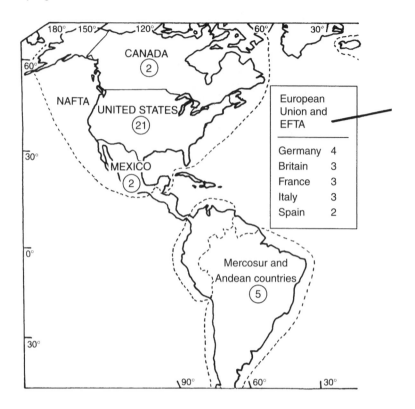

In 1991, Mercosur (Mercado Común del Cono Sur – Common Market of the Southern Cone) was created by Argentina, Brazil, Paraguay and Uruguay; trade among them had increased fourfold by the late 1990s. In a 1991 treaty, the Andean Group (Bolivia, Colombia, Ecuador, Peru and Venezuela) aimed to establish a common market within a few years. In 2004 the two groups announced plans for a continent-wide South American Community of Nations *(74).*

A similar target was set in 1990 by the Caribbean Common Market states (the ex-British West Indian islands, plus Belize and Guyana); and in 1993 the six-member Central American group undertook a gradual approach to a customs union. In South-East Asia, the ASEAN countries *(60)* agreed in 1991 to create a free-trade area; they began to reduce tariffs in 1993.

The membership of the OECD now overlaps with that of an even looser grouping, the Asia–Pacific Economic Co-operation forum (APEC). Created in 1989, it is by population the largest regional trade association. Since 1998 it has included all larger Pacific Rim economies – Australia, Canada, Chile, China, Japan, South Korea, Mexico, New Zealand, Papua New Guinea, Peru, Russia, Taiwan, the United States and the ASEAN countries (excluding Cambodia, Laos and Myanmar). The Americans and some others have been trying to encourage reductions of trade barriers between the members.

On a worldwide scale, important reductions of barriers to trade were achieved in 1994, with the signing of an agreement thrashed out during the complex negotiations called the 'Uruguay round'. A new World Trade Organization (WTO) replaced the 1948 General Agreement on Tariffs and Trade (GATT). A decade later almost all countries in the world were either members of the WTO or were 'observers' that planned to apply.

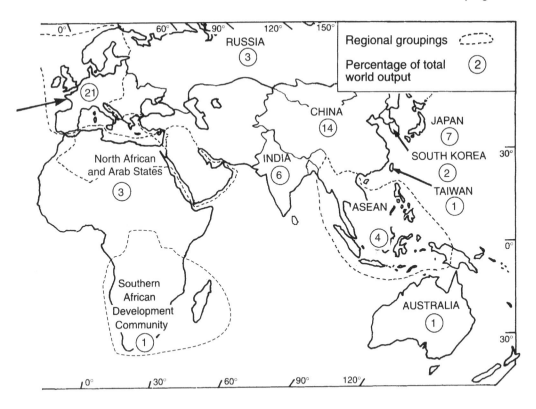

The initial goals of the WTO were to achieve reductions in tariffs and the European, Japanese and American farm subsidies which deprived other countries of markets and caused injurious 'dumping' of artificially created food surpluses. But huge subsidies continued, as did many restraints on trade – often less obvious, but more effective, than tariffs – and governments and lobbies showed great skill in preserving them. Further promotion of international trade has also been slowed by the anti-globalization movement, which sees consideration of workers' rights, environmental impacts and human rights as essential components of trade agreements. Globalization has, however, continued, with international trade growing faster than the world economy.

3 Energy

In the 1960s coal was succeeded as the world's biggest source of industrial energy by oil (petroleum) and the natural gas that is often found with it. World production of crude oil, which had been 275 million tons in 1938, rose to 1,050 million tons in 1960, to 2,275 million in 1970 and to 3,100 million in 1979.

The upsurge was then checked. Output fell by 10% between 1979 and 1981. This was a reaction to the startling increases in oil prices in the 1970s. Prices almost quadrupled in 1973–4 and then tripled in 1978–80. In both cases the rises were linked to turbulence in the Middle East.

Hardest hit were oil-importing countries in the 'poor south' (2), but recession and inflation also hit rich industrialized states. In many places, demand for energy stopped growing; in some, it fell. Where they could, consumers switched from oil to other sources of energy.

Just before the 'oil shocks', the pattern of production had been changing rapidly. In 1960,

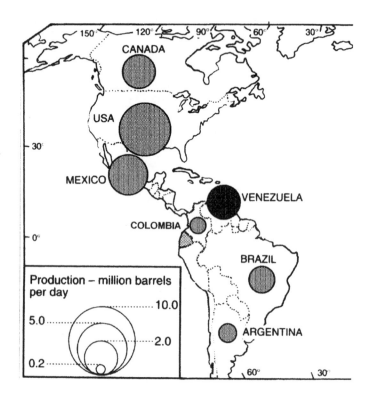

34% of all oil output was in the United States, only 25% in the Middle East and North Africa *(41)*. But Middle East oil was easy to extract and thus cheap. By 1970, Middle East and North African oil made up 40%, US oil only 21%, of a world output which had doubled in ten years. Europeans, Japanese and, to a lesser extent, Americans were heavily dependent on oil from Arab states and Iran.

The Organization of Petroleum-Exporting Countries (OPEC) was founded in 1960, and the members of this 'cartel' began to try to raise prices, mainly by limiting output. (The present members of OPEC are Algeria, Indonesia, Iran, Iraq, Kuwait, Libya, Nigeria, Qatar, Saudi Arabia, United Arab Emirates and Venezuela.) Their strength was suddenly increased during the 1973 Arab–Israeli war *(43)*, when Arab states cut off supplies to some western countries. Prices soared, while economies sagged – except in the oil-exporting states. The second big wave of price rises began in 1978, when Iran's exports were sharply reduced during the last turbulent months of the Shah's rule *(47)*.

By 1981 the recession caused by this 'oil shock' had cut demand enough to start prices falling. In the 1980s the Iran–Iraq war had little effect on total output or on prices; the price rises of the 1970s had encouraged exploration and production in non-OPEC countries, including Britain, whose North Sea fields *(22)* soon made it a net exporter. A steep price fall in 1986 followed Saudi Arabia's decision to raise its output, after keeping it at only a third of capacity for several years; the Saudis, the biggest OPEC producers, saw that the cartel's strategy of raising prices had boomeranged, bringing it new competitors, and they reverted to exploiting their ability to produce oil cheaply. By the late 1980s oil prices were lower, in real terms, than they had been in the mid-1970s.

In 1990 there was only a brief jump in prices when Iraq seized Kuwait. For a year after that, no oil came out of Kuwait; for the next six years, only a trickle came out of Iraq, which

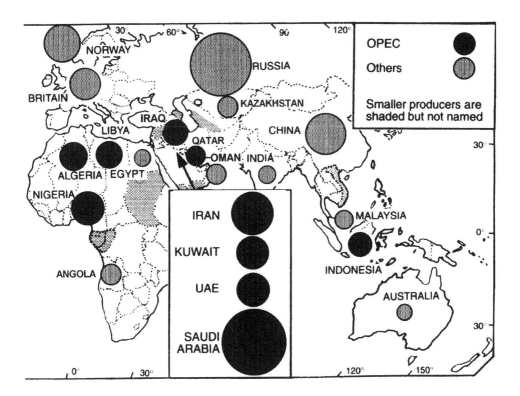

had brought a United Nations embargo upon itself. (Until 1996, Iraq rejected UN offers to let it sell a limited amount of oil on condition that part of the proceeds should go to compensate the victims of its actions – *48*.) Another big increase in Saudi output helped to fill the gap. After declining to below production cost for some non-OPEC producers in the late 1990s, prices stabilized for a time. With the US action against Iraq in 2003 and increasing demand from China, prices have recently risen steadily, beginning to approach (in real terms) the late-1970s high.

World demand remained at much the same level from 1980, when the great oil upsurge had ended, through the mid-1990s. Since then improvements in energy efficiency have not matched economic growth.

Over 60% of known oil reserves are in the Middle East, and Saudi Arabia alone has almost 25%. Iraq, Iran and Kuwait have about 10% each. If oil sands (oil-saturated sand deposits that are more expensive to process than conventional liquid crude oil) are included, Canada has about 15% of world reserves. In the early 2000s the OPEC members' annual output was around 10 billion barrels; their share of total output is 40%.

The map shows the OPEC states and other large producers. Smaller non-OPEC producers include Azerbaijan, Brunei, Congo-Brazzaville, Denmark, Ecuador, Equatorial Guinea, Gabon, Sudan, Syria, Turkmenistan, Vietnam and Yemen *(6)*. Some producers are net importers; most conspicuously, the United States consumes a quarter of total world output, but produces only an eighth.

Natural gas has, in recent years, provided an amount of energy roughly equivalent to 60% of all the oil produced. It is estimated that around 30% of known reserves of gas are in Russia. At least 40% more are in the Middle East. Until the 1960s gas was mostly piped only to places in its country of origin; now long pipelines take it from Russia to both eastern and western Europe and from the Caspian Sea region to the Mediterranean through Turkey. Tankers carry liquefied natural gas on long sea routes, notably from Australia and Indonesia to Japan. Much gas is still flared (burned off, therefore wasted) as a by-product of oil production.

The US has more than 25% of the world's coal reserves, with Russia, China, India and Australia together accounting for another half. Concerns about air pollution and the difficulty of adapting coal for use in vehicles prevent it from replacing oil in oil-importing countries. High oil and gas prices could encourage greater use of coal in power generation.

The nuclear power plants built since the 1950s now provide 16% of the world's electricity (6% of its total energy supplies). Some countries' nuclear reactors provide a much bigger proportion of their electricity: 75% in France, 50% in Sweden and Ukraine, 40% in South Korea, 30% in Germany, Hungary and Japan.

All this power has come from uranium. (To use lighter elements would require the harnessing of controlled processes of nuclear fusion; that has not yet been achieved, despite considerable research spending.) Uranium is found in many places, but America, Australia, Brazil, Canada, Kazakhstan, Namibia, Niger, Russia and South Africa together have more than 80% of recoverable reserves. Canada and Australia now mine half the world's uranium, and most of the rest is produced in the former Soviet Union (Kazakhstan, Russia, Uzbekistan) and Africa (Niger, Namibia) *(4)*.

In some countries there has been sharp controversy about nuclear energy. Its advocates point out that coal and oil cause pollution, may cause 'global warming', and will in time be exhausted *(1)*. Its opponents emphasize the risks: their arguments were underlined by the disaster in 1986 at the Chernobyl plant in Ukraine *(18)*, which spread radioactivity across Europe. The disposal of dangerous reactor waste remains an unsolved problem. Falling prices of coal and oil helped to reduce enthusiasm for nuclear energy in the 1980s and early 1990s, but more recently high oil prices have begun to reawaken interest in it.

The use of nuclear power will be encouraged by continued rapid economic growth, especially in Asia, and consequent competition for oil. At present, China and India generate less than 3% of their electricity from uranium, and both are heavily dependent on oil imports. World hydroelectric capacity is unlikely to increase much, because of growing opposition to dam-building on environmental grounds. Use of wind, solar and other renewable power sources is expected to expand rapidly, but these together account for only 1% cnt; of global energy production. Increased efforts at conservation will help slow the rate of growth in the demand for energy but are unlikely to produce an actual reduction in world energy needs.

4 Nuclear Geography

Nuclear weapons inspire a natural horror; yet their influence on world affairs over the past half-century was not entirely evil. The two bombs dropped on Japan in 1945 saved far more (mainly Japanese) lives than they took. They forced Japan's ruling militarists to abandon their plans to make the country fight to its last woman and child against the coming Allied invasion – which had been seen as the only means of ending the 1939–45 war and liberating all the Japanese-occupied lands. And in the subsequent decades, at several critical moments, the fear of nuclear devastation helped to prevent tense east–west confrontations from exploding into full-scale war.

But the two superpowers' efforts to maintain an uneasy 'nuclear balance' – plus the creation of the relatively small British, Chinese and French nuclear arsenals – burdened the world with an almost unimaginably large array of powerful new weapons. And the old

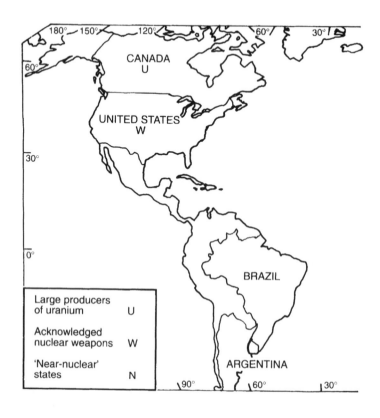

concepts of war were transformed. Instead of troops marching across a frontier, nuclear missiles could now be sent to strike an enemy thousands of miles away.

During the 1960s both the Soviet Union and the United States developed and deployed intercontinental ballistic missiles (ICBMs) whose range of over 6,000 miles made it possible for them to strike each other across the Arctic *(77)*. Both of them sent missile-firing submarines out into the oceans (Britain and France soon followed this example). The Soviet arsenal included intermediate-range ballistic missiles (IRBMs), capable of hitting western Europe, China or Japan; China was not slow to respond by targeting Soviet cities with IRBMs (by the 1980s it was deploying ICBMs). As well as ballistic missiles (high-trajectory missiles moving at bullet-like speed), America built 'cruise' missiles – sophisticated developments of the German V-1 'pilotless planes' used against London in 1944 – which could find their targets by map-reading.

The new Soviet leadership installed in 1985, headed by Mikhail Gorbachev, saw that the floundering Soviet economy had to be relieved of the ever-growing burden of the arms race. In the first east–west disarmament deal, the 1987 INF (intermediate-range nuclear forces) treaty, the two superpowers undertook to destroy 2,600 missiles. In the 1990 CFE (conventional forces in Europe) treaty, the 22 NATO and Warsaw Pact governments *(10)* agreed on phased reductions which would leave the two alliances with forces of roughly equal size in the area between the Atlantic and the Urals. (The subsequent disappearance of the Soviet Union and of the Warsaw Pact left the CFE treaty in need of much revision, but its essence was preserved.) There followed the START-1 (strategic arms reduction) treaty, committing America and Russia to reducing the number of their strategic missiles

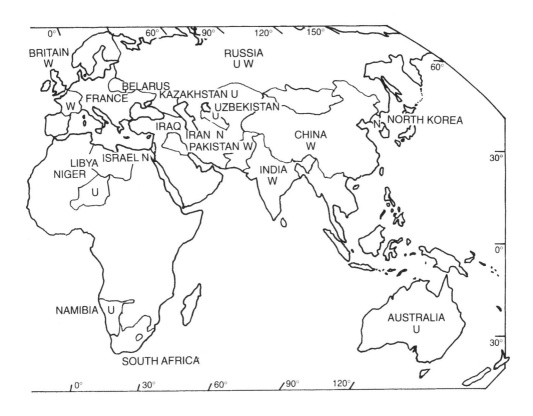

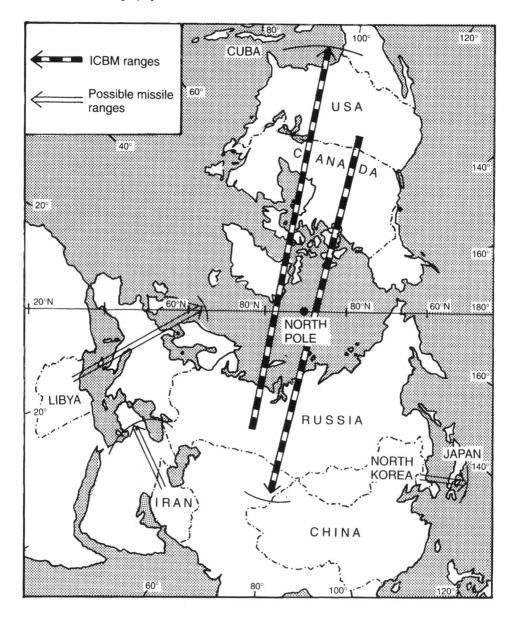

and long-range bomber aircraft to about 6,500 on each side by the year 2001. Implementation of its successor, START-2, was complicated by US withdrawal from the 1972 anti-ballistic-missile treaty to allow for development of missile defence systems. However, there followed the SORT (strategic offensive reductions) treaty, committing each side to a maximum of 2,200 warheads by 2012. Destruction of surplus nuclear warheads continued, with financial support from the US to aid Russia in securing its nuclear facilities.

After China's first test explosion, in 1964, there were five recognized nuclear-armed nations: America, Britain, China, France and the Soviet Union. The 1968 Nuclear Non-Proliferation Treaty (NPT) was designed to prevent an expansion of the 'nuclear club'. Members of the 'club' were to undertake not to help non-members to acquire nuclear arms.

Other signatories, if they had nuclear power plants, were to allow inspectors from the International Atomic Energy Agency (IAEA) to ensure that no material was diverted into making weapons.

All nuclear reactors produce plutonium, which can be used to make atomic bombs – but only after reprocessing. Bombs can also be made with highly enriched uranium. However, both methods require the construction of special installations. By 1996 most of the countries that had the necessary advanced technology had signed the NPT and accepted IAEA 'safeguards'; nuclear-armed China and France, after long resistance to the NPT, had accepted it. Argentina, Brazil and South Africa had abandoned their attempts to make atomic bombs; in 1992, Argentina and Brazil accepted the 1967 Tlatelolco Treaty *(74)*. But there was still cause for concern about some countries that were rated as 'near-nuclear'.

In the 1990s, India, Israel and Pakistan were still rejecting the NPT, and (despite official denials) it was believed that they had made much progress towards acquiring nuclear arsenals. India had staged a test explosion in 1974, claiming that it had only 'peaceful purposes' in mind, and carried out further tests in 1998. These prompted Pakistan to conduct its own tests the same year, expanding formal membership of the 'nuclear club' to seven. In 2006 the US and France signed agreements with India allowing for the transfer of 'civilian' nuclear technology. India accepted IAEA inspection of 'civilian' nuclear facilities – but it did not agree to limit future weapon production, allow monitoring of 'military' facilities, or sign the NPT. Israel is not known to have carried out tests, but it is thought that Israel has completed the bomb-making process and has constructed a number of weapons in an effort to discourage attack by its larger neighbours.

The breaking up of the USSR in 1991 temporarily increased the 'nuclear club' membership: some of the former Soviet nuclear weapons were in Belarus, Kazakhstan and Ukraine. However, with American financial and technical help, all the warheads had been transferred to Russia by 1996. The larger effort to identify and secure (or destroy) all nuclear, chemical and biological weapons materials in the former USSR has made considerable progress but is likely to take several decades to complete.

Three countries that had signed the NPT failed to give the IAEA inspectors full access to their nuclear installations: Iran, Iraq and North Korea (the US-designated 'axis of evil'). From 1991 until the American invasion in 2003, a United Nations commission was unearthing and destroying Iraq's facilities for making nuclear, chemical and biological weaponry *(48)*. North Korea, which had signed the NPT in 1985, withdrew in 1993, then suspended its withdrawal, but continued to make difficulties about accepting inspection. After ejecting inspectors, North Korea once again withdrew from the NPT in 2003, claiming in 2005 to have constructed nuclear weapons *(59)*. Iran's nuclear programmes reached a crisis after 2003. It alternately allowed access by inspectors and refused to co-operate with the IAEA, while insisting that its uranium-enrichment programme was for peaceful purposes. Like Libya – which abandoned its nuclear weapons programme in 2003 – Iran received nuclear technology and material from Pakistan. Attempts by European countries to negotiate an end to Iran's pursuit of nuclear weapons had not succeeded by 2006.

At NPT review conferences, the 'club' members were repeatedly warned that the non-proliferation system would break down if they did not fulfil their promises to cut back their development of new weapons technology or 'vertical proliferation' – as distinct from the 'horizontal' proliferation (cross-border technology transfer) banned by the NPT. As a priority, they were urged to stop nuclear tests. A 1963 treaty had limited America, Britain and the USSR to underground tests (France's testing was underground from 1974, China's from 1980). In 1996 a comprehensive test ban treaty was at last concluded, and signed by most governments, including those of America, Britain, China, France and Russia. A similar treaty banning the production and use of chemical weapons was adopted in 1993 and ratified by most countries.

As the huge arsenals of the nuclear-armed powers began at last to shrink, more attention was turned to the growing number of missiles in the hands of 'Third World' states *(6)*. During Iraq's wars, it fired ballistic missiles at Tehran in Iran, at Tel Aviv in Israel and at Riyadh in Saudi Arabia *(48)*. About thirty Third World countries have ballistic missiles, including Algeria, Egypt, India, Iran, Libya, Pakistan and Syria. There was particular concern about North Korea, not only because of its rulers' unpredictable and secretive ways but also because it was known to have missiles with a 700-mile range and to be developing ones with a range of more than 3,000 miles. Missile technology from China, North Korea and Russia is believed to have supported missile development in Iran, Libya, Pakistan and Syria. North Korea and Iran combined possession of missiles with an alarming eagerness to acquire nuclear weapons *(47, 59)*. And, although only a few of the missile-possessing states were also classed as 'near-nuclear', concern was not limited to these cases. In the 1991 'Gulf war' there had been fears that the ballistic missiles fired by Iraq might prove to be carrying chemical or biological warheads.

5 Sea Law

Seven-tenths of the earth's surface is covered by the 'seven seas'. Until recently, nearly all of this vast area was under no national jurisdiction. Most coastal states claimed territorial waters extending only 3 nautical miles from shore (100 nautical miles are about 115 land miles or 185 kilometres). But some states' claims grew larger and larger as expanding populations and industries increased the demand for fish and oil (petroleum). Disputes over fishing rights, and rights to seabed oil, became more frequent. United Nations conferences on the law of the sea (UNCLOS) in 1958 and 1960 failed to resolve most of these problems. However, the 1958 conference produced a convention on the 'continental shelf' (the relatively shallow offshore part of the seabed); it was on this basis that the countries around the North Sea shared the rights to the oil and gas lying beneath it *(22)*.

The third UNCLOS ran from 1974 to 1982 and negotiated a whole new code of sea law. The standard for territorial waters was set at 12 miles, and each coastal state would also control fishing and extraction of seabed oil in a 200-mile-wide 'exclusive economic zone' (EEZ). A third of all the oceans was thus to come within the jurisdiction of coastal states. Even a small island state could claim a sea zone of some 130,000 square miles. Where two states' zones overlapped, a median line, equidistant from their coasts, would normally be drawn.

The general right of freedom of navigation was upheld in the EEZs, and also in straits of international importance, even where these straits became territorial waters (for example: with 12-mile limits, any ship passing through the 21-mile-wide Dover strait must enter British or French territorial waters). The new code did not affect the existing special rules that apply to the Turkish straits (Bosporus and Dardanelles) or those that apply to the Suez and Panama canals.

The UNCLOS code needed to be ratified by at least sixty nations, and this was not achieved until 1994, when it duly came into force. One cause of delay was a long wrangle over the code's provisions about future 'mining' on the deep ocean floor. The United States and Germany, in particular, objected to these clauses, arguing that they would penalize those states that pioneered new methods of dredging up minerals from the depths; eventually some amendments were agreed.

Long before 1994, many states had acted as if the new code were already in force, claiming EEZs and often getting into disputes based on those claims. In some cases, long-dormant disputes over small islands were intensified when it was seen that claims to large areas of sea might now be involved; this was relevant to Argentina's quarrel with Chile about the Beagle Channel islets and to its 1982 attempt to annex the Falklands *(75)*.

By adopting the new code, the world's governments may well have averted a general collapse of sea law into anarchy. They have not necessarily averted a collapse of the

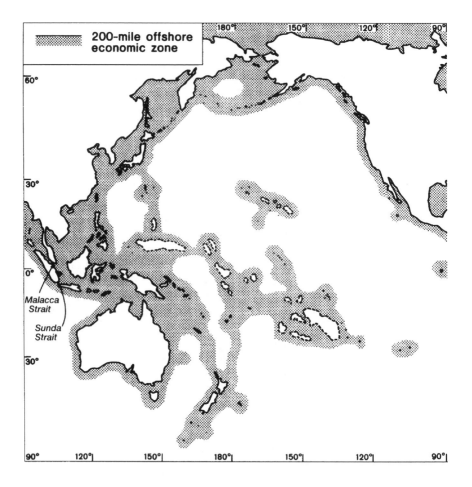

world's fisheries. Between 1950 and 1990, the total world catch showed a fivefold increase. In 1995 the UN Food and Agriculture Organization (FAO) gave warning that nine of the world's 17 major fishing grounds had already been disastrously overfished, and that four others looked like sharing that fate.

Nine-tenths of the world's fish stocks are in waters less than 200 miles from shore; now that the coastal states have full powers of control in their EEZs, they ought to be able to ensure the conservation of those stocks. A UN agreement on fish stocks, ratified in 2001, calls for regional co-operation in the management of migratory fish species, including conservation measures where necessary. It has been signed by most large Atlantic but not many Pacific fishing nations. But many states, while taking action against foreign intruders, failed to prevent overfishing by their own fleets and indeed encouraged it by granting lavish subsidies. Some small states sold fishery rights in their waters to bigger nations, whose fleets then fished out those waters with no regard for conservation. And sharp disputes still arose about the exact positions of vessels fishing near to the 200-mile zones, especially when they were equipped to make huge catches. In 1996, for example, Ireland accused Japanese ships of entering its zone, each towing a fishing line 70 miles long.

The losses caused by unsustainable fish harvests are compounded by the large fraction of 'by-catch' or unwanted fish, dolphins and turtles taken, then discarded, as part of

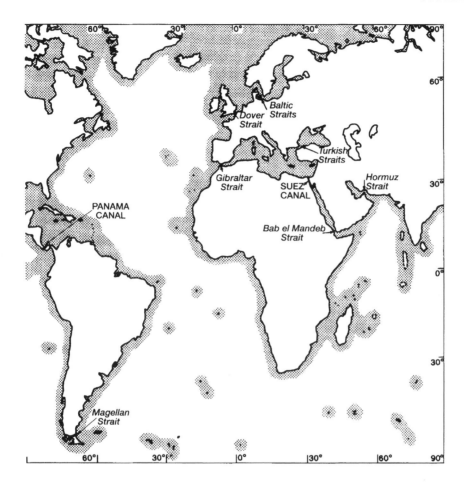

commercial operations. The result is a continuing decline in fish populations worldwide. A corresponding increase in aquaculture (fish-farming) makes up part of the loss but introduces new problems of pollution and damage to coastal ecology.

Most nations adhere to the 1986 international moratorium on commercial whaling. Japan, Norway and Iceland continue to hunt whales, claiming that some whale species are no longer threatened with extinction and that whales eat fish needed for human consumption.

6 No Longer Three Worlds

By the early 1990s the world could no longer be seen as divided, like Caesar's Gaul, into three parts. That division had taken shape between the late 1940s and the 1960s. The Soviet Union had sealed off its newly enlarged domains behind its 'iron curtain', and joined hands with a China whose civil war had brought it under communist rule, to create a formidable-looking 'east'. Fear of Soviet armed strength led most of the west European states to form, with the United States and Canada, the largest alliance that had ever existed in peacetime; with a few democratic 'neutrals', they constituted the 'west' – not a monolithic bloc, but a group which, under new pressures, was showing a new unity. Meanwhile, from 1946 onwards, decolonization in Asia and Africa (27) was creating new independent states whose basic interests were those of the 'south', or the 'Third World'. East–west issues were of little

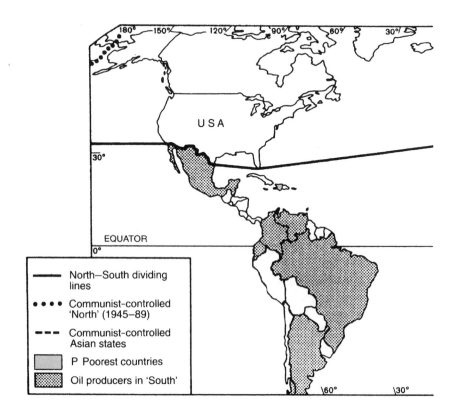

interest to their new rulers. Indeed, at times it seemed that, in third-world eyes, there were only two worlds: south and north.

These terms were not geographically exact. Japan was often lumped in with 'the west'; Australia and New Zealand were labelled as 'north'; China might be called 'east' or 'south'. But it was broadly true that west and east, between them, had most of the world's economic power and most of its armaments; and that the south was poor, preoccupied with the kind of difficulties that new nations face (including many local quarrels), and reluctant to get involved in east–west conflicts. An embodiment of this reluctance, the 'non-aligned' movement, launched in Belgrade in 1961, eventually drew in 113 member states.

Some members of that movement were not truly non-aligned; they received subsidies from, or had defence links with, either the Soviet Union or western powers. And alignments became still more complex after the Soviet–Chinese breach in the 1960s. The 'oil shocks' of the 1970s *(3)* further divided the third-world states, enriching some and impoverishing others; later, gaps widened between states with fast-growing economies and those that were making no progress at all. But a basic three-worlds pattern remained visible until the end of the 1980s brought the ending of Soviet domination in eastern Europe, followed by the break-up of the Soviet Union itself.

Communist rule continued in China, North Korea, Vietnam, Laos, Cambodia (until 1993) and Cuba, and for a time in some parts of eastern Europe and of the former Soviet Union. To that extent, it might be said that there were still three worlds; or perhaps two and a half. And there were signs of some Russians' aspirations to reassemble the Soviet Union (or part of it), in substance if not in form. But during the early 1990s, in most parts of what had once

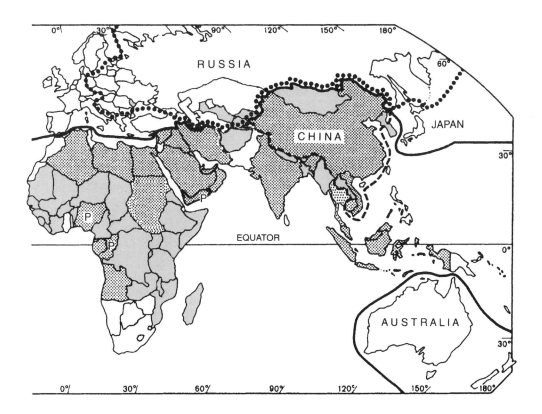

been a solid communist eastern bloc, the economic structures were becoming more open, and so were the frontiers. East and west were no longer two antagonistic rival camps, maintaining a precarious military balance and defending totally opposed economic and political systems. When antagonism did emerge, national factors were more likely to underlie it than ideological ones. For example: when, in the turmoil of what had been Yugoslavia, Russia opposed some western moves to restrain the Serbs, the main reason was that Serbia had old historical claims to Russia's sympathy.

The breakup of the 'east', which national factors had hastened, strengthened those same factors in both 'west' and 'south'. Several members of the 'south', among them South Korea, Taiwan, Hong Kong, Israel and Singapore, built technologically advanced economies and achieved income levels exceeding those of some 'northern' countries. China's communist government oversaw the development of a capitalist economy rivalling that of Japan. The western allies, feeling less need to hold together, bickered about Yugoslavia, Cuba, Iran, Iraq and Libya, about the future of the European Union, and about admitting east European applicants to the union or the alliance. In the Third World, the loss of Soviet subsidies, arms supplies and diplomatic support brought some rulers down and made others change course. Some regimes which had long counted on American support, because they took anti-communist attitudes, found the Americans becoming less charitable about some of their other actions. And those who had tried to play east and west against each other learned that this particular game was over. The three-worlds framework, so restricting and yet so reassuringly familiar, was gone.

7 United Nations

Today there are more than 190 sovereign states; 80 years ago there were only 70, but many empires have broken up – most recently the Soviet empire. Nations and communities that had been parts of an empire have quarrelled, sometimes going to war (e.g. Armenia and Azerbaijan, India and Pakistan, Cyprus, Congo, Sudan). And the chances of war breaking out have been increased simply by the increase in the number of nations.

Nationalism has become a prevalent force. But internationalism has also developed. The United Nations, the world organization created after the 1939–45 war, has lasted more than twice as long as the League of Nations, its predecessor. America never joined the League, and Germany and the Soviet Union were not members for long; but today the only significant absentee from the UN is Taiwan, prevented from joining by China.

Operations by UN forces, UN observer missions, and multinational forces acting with UN authorization (some entries include multiple operations)

(a) *Ended by 2005*
1 Indonesia, 1947–51 *(64)*
2 Greece, 1947, 1952–4 *(15)*
3 Korea, 1950–3 *(59)*
4 Lebanon/Syria, 1958
5 West New Guinea (Irian), 1962–3 *(64)*
6 Yemen, 1963–4 *(46)*
7 Dominican Republic, 1965–6 *(71)*
8 Egypt/Israel, 1956–67, 1973–9 *(43)*
9 Afghanistan/Pakistan, 1988–90 *(50)*
10 Iran/Iraq, 1988–91 *(48)*
11 Angola, 1988–99 *(33)*
12 Namibia, 1989–90 *(33)*
13 Central America, 1989–92 *(71)*
14 Cambodia, 1991–3 *(62)*
15 El Salvador, 1991–95 *(71)*
16 Iraq/Kuwait, 1991–2003 *(48)*
17 Mozambique, 1992–4 *(32)*
18 Somalia, 1992–5 *(35)*
19 Former Yugoslavia, 1992–2002 *(15)*
20 Uganda/Rwanda, 1993–4 *(36)*
21 Rwanda, 1993–6 *(36)*
22 Chad/Libya, 1994 *(38)*
23 Tajikistan, 1994–2000 *(20)*

24 Croatia, 1995–8 *(15)*
25 Macedonia, 1995–9 *(15)*
26 Bosnia, 1995–2002 *(15)*
27 Guatemala, 1997 *(71)*
28 Central African Republic, 1998–2000
29 Sierra Leone, 1998–2005 *(37)*
30 East Timor, 1999–2005 *(64)*

(b) *Continuing in 2006*
31 Arabs/Israel, 1948– *(43, 44)*
32 India/Pakistan, 1949– *(51)*
33 Cyprus, 1964– *(26)*
34 Syria/Israel, 1974– *(43, 44)*
35 Lebanon, 1978– *(43, 45)*
36 Western Sahara, 1991– *(40)*
37 Haiti, 1993–2000, 2004– *(71)*
38 Georgia, 1993– *(19)*
39 Kosovo, 1999– *(15)*
40 Congo (Zaire), 1960–4, 1999– *(32, 36)*
41 Ethiopia/Eritrea, 2000– *(35)*
42 Liberia, 1993–7, 2003– *(37)*
43 Ivory Coast, 2004– *(37)*
44 Burundi, 2004– *(36)*
45 Sudan, 2005– *(35)*

The UN today has 191 members, of which five – America, Britain, China, France and Russia (replacing the Soviet Union) – are permanent, veto-wielding members of the Security Council. Ten more Council members are elected for two-year terms. Financial contributions are levied on the basis of economic strength, with the US expected to contribute 22%, Japan nearly 20%, and the 25-member EU 38%. Voting is on the basis of one country, one vote, so the weight of India (population 1.1 billion) equals that of Tuvalu (12,000). Recent calls for reform of Security Council membership, to involve expansion or addition of new permanent members, are tempered by national rivalries and are therefore unlikely to succeed in the near future.

The work of the 'UN family' on such matters as trade, food, aviation, refugees, drugs and health has mostly been done through its 'specialized agencies'. The primary role assigned to the United Nations itself was to give governments a means of working together to prevent wars, or at least to limit their effect. Sometimes diplomatic or other pressures, such as economic 'sanctions', might suffice. But the new practices evolved in the UN's six decades of activity have often required the use of soldiers, in peacekeeping forces or as observers. Experience has shown that 'peace is too serious to be left to civilians'.

Operations of this kind became more numerous from the late 1980s onward. Relations between the major powers were improving, so a proposal to take action through the UN was less likely to be vetoed by one of the five permanent members of the Security Council. There was also growing concern about stopping civil wars – which, in recent years, have greatly outnumbered international ones. Indeed, the only active armed conflicts between

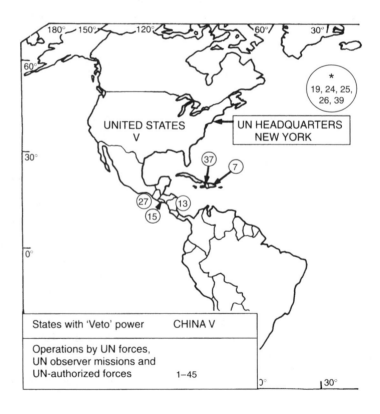

sovereign states are minor affairs by historical standards, and there is no immediate potential for any major wars – a rare situation in centuries of human history.

There has been only one instance of a UN member state invading and annexing another
member state: Iraq's seizing of Kuwait in 1990. Iraqi obduracy made it impossible to
negotiate a withdrawal which a small UN force or observer mission could supervise. To
liberate Kuwait, a full-scale military operation was necessary. The American-led forces that
carried it out were not UN forces, but there was enough agreement on the need for action to
enable the Security Council to authorize it.

The UN's founders designed it to tackle conflicts between states, but not conflicts inside
them. At times, this distinction could be hard to make. In some civil wars (e.g. in Greece,
1945–8, or Angola, 1975–91) there was blatant meddling by other countries; in some there
was a danger of other countries being drawn in (e.g. Cyprus, 1961–4). But the fighting
from 1991 to 1995 in Somalia was a purely internal conflict; nothing could be done to halt
or limit the war by deterring other countries from taking a hand (nor could peace terms be
settled with Somalia's own government, as there was none). By the mid-1990s, experiences
such as this were making UN member governments more cautious about trying to stop
civil wars.

The shift from international to internal conflict has left the UN open to criticism for
inaction. It did not, or was not permitted to, intervene to prevent the Rwandan genocide
of 1994 or recent wars in Sudan. On the international level, its ability to resolve Middle
East conflicts is limited by the refusal of some members to recognize Israel (which has
never been permitted to serve on the Security Council). Credibility in other areas has

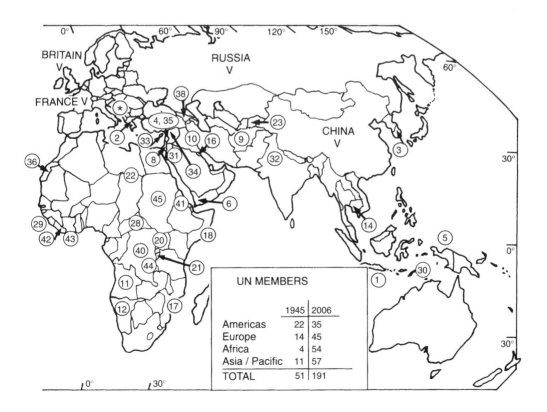

been harmed by financial crises, and by the appointment of countries with poor human rights records to the Commission on Human Rights (including Libya as chair in 2003). Current reform efforts may enhance the UN's ability to act against genocide and terrorism.

8 Terrorism

Politically motivated violence against civilians is not new, but its projection across national borders has increased in recent decades. This transition is marked by the emergence of the al-Qaeda terrorist network, whose first large-scale attack was the 1998 bombing of American embassies in Kenya and Tanzania.

Since then, 90% of terrorist attacks and casualties have occurred in the countries shown on the map. The more deadly recent attacks reflect the activities of Islamists – including the largest attack, when members of al-Qaeda hijacked aeroplanes for suicide attacks on New York and Washington that killed 3,000 people on 9/11 (September 11, 2001). The largest area of recent terrorist activity has been Iraq after the US overthrew its dictator in 2003 *(48)*; Afghanistan *(50)* also ranks high. Outside Iraq and the USA, the deadliest individual attacks were committed by Islamists in India, Indonesia, the Philippines, Russia and Spain; and by non-Islamic rebel groups in Angola and Uganda. Frequent, if somewhat less deadly, attacks have occurred along the borders of the Muslim world *(28)* – in the Caucasus, Israel, Kashmir, the Philippines and southern Thailand – and, reflecting local rebellions, in Colombia, eastern India, Nepal and Sri Lanka.

There is no agreement on the limits of 'terrorism'. Individual incidents are generally of short duration, directed at non-military targets that are sometimes randomly selected, and designed to instill fear rather than achieve specific military or political goals – though political changes are often sought. Thus, genocides, revolutions, coups and military conflicts are not classed as terrorism, though terrorist action may be associated with them. An armed rebellion against a government and its army, as in the conflict between UNITA and the Angolan government *(33)*, may involve both 'war' and terrorist incidents.

Repressive regimes suffer less terrorism than democracies because they can crush dissent more vigorously and control the movement of their people. Thus, of the states on the map, only Pakistan is clearly under military or dictatorial rule – and government control of the country is not complete there. Authoritarian governments are, however, vital supporters of terrorist groups outside their borders. Iran and Syria arm and finance Hezbollah, Hamas and other Palestinian terrorist groups, and until recently Iraq and Libya did the same; Sudan and Afghanistan hosted al-Qaeda in the 1990s; India accuses Pakistan of failing to control Pakistan-based terrorist groups responsible for bombings in India; Cuba has aided Colombian rebels. Western support for violent rebel groups or dictators has also contributed to terrorist activity; for example, the US supported Central American armies and guerrillas that killed civilians during the 1990s.

As the frequency of interstate wars has declined, armies designed to fight each other have had problems adapting to more amorphous challenges. Terrorist organizations are rarely conquered in war; 'wars on terror' mostly kill the civilians who live where terrorists

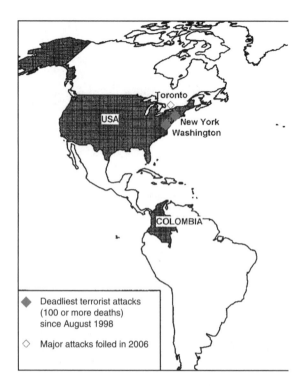

Deadliest terrorist attacks
(100 or more deaths)
since August 1998

Major attacks foiled in 2006

operate and are used as their 'human shields'. Combined political, popular and military pressure can encourage negotiation. Several terrorist organizations reduced or ended their activities in the 1990s and 2000s, including the Irish Republican Army, the Kurdish PKK in Turkey, and some of the Colombian paramilitaries *(72)*. Other groups – such as Hezbollah in Lebanon – increased in power, using external sources of money and arms to resist state or popular opposition.

Prevention of terrorism depends on intelligence-gathering and security measures. Apparent schemes to blow up multiple aeroplanes leaving London in 2006 and airports in

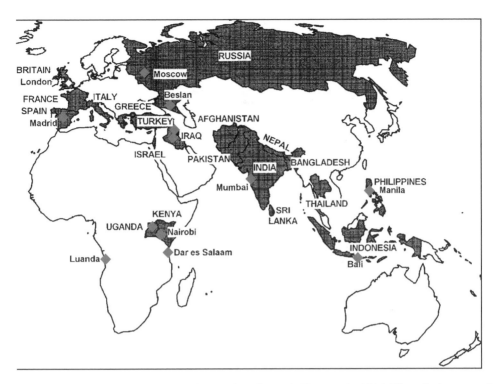

Asia in 1995 were foiled, as were planned bombings in Toronto in 2006. Though threatened by numerous terrorist groups, Israel has deterred hijackings with tight airport security and reduced the level of land-based violence through construction of border fences.

Apart from a gas attack on the Tokyo subway in 1995, terrorists have relied on conventional weapons. The September 2001 attack on the US was carried out by hijackers with knives. But the methods and materials for creating 'weapons of mass destruction' are readily available, and no means has yet been devised to scan all travellers and cargo shipments; attacks that kill thousands rather than dozens can only become more common.

9 Commonwealth

The Commonwealth consists almost entirely of countries which were once part of the British empire. (It is a voluntary association; Burma, on becoming independent, chose not to join.) The old Dominions were confirmed in their independence by the 1931 Statute of Westminster. India and Pakistan became independent in 1947. The 'decolonization' of other dependencies followed, and by 1965 the Commonwealth member states numbered 22; by 1985, 49; and, in 2006, 53.

The former Irish Free State left the Commonwealth on becoming a republic in 1949 (retaining some special privileges; for instance, Irish citizens living in Britain can still vote in British elections). South Africa left in 1961 but was readmitted in 1994; Pakistan left in 1972 but returned in 1989. Fiji's membership was suspended from 1987 to 1997, from 2000 to 2001, and again in 2006, Nigeria's in 1995–9, and Pakistan's from 1999 to 2004, in each case following the overthrow of an elected government or human rights violations. After

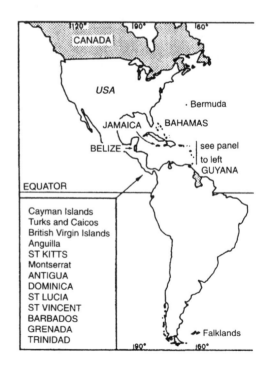

criticism of its presidential election in 2002 *(32)*, Zimbabwe was suspended and withdrew from the Commonwealth the following year.

Membership was not contemplated for some states, such as Sudan, which, although under British control for a time, had not been formally classed as British dependencies. On the other hand, membership was granted in 1980 to Vanuatu, formerly a British–French condominium; and in 1995 to Cameroon, the larger part of which had been under French rule. These two, and other Commonwealth states, including Canada, also joined the *'Francophonie'* association, many of whose 47 members had been French dependencies. Similarly, Mozambique, formerly a Portuguese dependency and part of the eight-member Community of Portuguese Language Countries, was admitted to the Commonwealth in 1995 'as an exceptional case'.

One very small state, Nauru *(66)*, is called a 'special member'; it does not take part in the 'summit' meetings of heads of government (prime ministers and presidents) that are normally held every two years (Tuvalu, formerly a special member, became a full member in 2000). Naturally, the smaller and poorer members have benefited most from the services of the Secretariat, the technical co-operation fund and other joint institutions; but all members have seen the advantages of remaining in a group that bridges the gap between the world's 'north' and 'south' *(2, 6)*.

Predictions that Britain's entry into the European Community in 1973 would destroy the Commonwealth proved wrong. It has been able to adapt itself to the fact that nearly all its members have also joined regional groupings and alliances. It has survived bitter disputes, and even wars, between its members.

In some of them, military rulers or one-party regimes have, at times, suppressed all opposition and flouted basic human rights. In such cases the Commonwealth, while

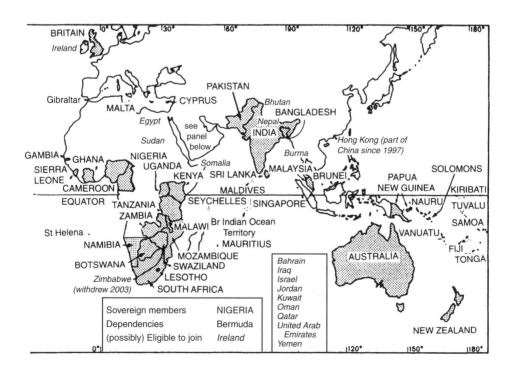

respecting each state's sovereignty, has tried to help restore freedom; for example, between 1990 and 2004 it sent 46 teams of observers to 26 member countries to improve the chances of elections being free and fair. The suspensions of Fiji *(66)*, Nigeria *(37)* and Zimbabwe were moves intended to encourage a return to democracy in those countries.

10 Europe: East and West

In the years 1989–91 the face of Europe was changed. This was the end of a 45-year period during which the continent had been more sharply divided into two parts than ever before (except in wartime). Eastern Europe's rulers sealed it off, depriving its inhabitants of contact with their western neighbours. Families were divided; letters were censored; western broadcasts were jammed; people who tried to escape to the west were killed.

In 1944–5, when Nazi Germany's hold on Europe was broken *(13)*, the advancing Soviet army imposed Soviet control on East Germany, Poland, Hungary, Romania and Bulgaria; and in 1948 a Soviet-backed communist coup turned Czechoslovakia into one more police state. An 'iron curtain' (in Winston Churchill's words) fell across the middle of Europe. The USSR made formal alliances with its satellite states, whose armies were, in practice, already under Soviet control; in 1955 a new treaty, the Warsaw Pact, consolidated these alliances. The Soviet economic grip on the satellites was formalized by the creation in 1949 of the Council for Mutual Economic Assistance ('Comecon'). The Hungarians made a bid for liberation in 1956; the Czechoslovaks tried, in 1968, to make their communist regime more tolerable; the Soviet army crushed both attempts, reimposing the 'iron curtain' *(14)*.

In western Europe (where, after 1945, some communist parties, and notably those of France and Italy, were strong), fear of Soviet domination produced in 1948 the five-member Brussels Treaty, and in 1949 the North Atlantic Treaty, which by 1982 had developed into an alliance of 16 democracies, including the United States and Canada *(11)*. Five European democracies, Austria, Finland, Ireland, Sweden and Switzerland, although 'western' in their sympathies, remained 'neutral' for special reasons, and two communist states, Yugoslavia and Albania, broke away from the 'Soviet bloc' *(15)*; but these were minor exceptions to the basic rule of east–west division in the Europe of 1945–89.

The Soviet rulers repeatedly called for all-European 'security' agreements, their main aim being to legitimize the partition of Germany and the communist hold on eastern Europe. What eventually emerged suited them less well. At Helsinki in 1975, after long negotiations in the Conference on Security and Co-operation in Europe (CSCE), an agreement was signed by thirty-three European governments, and by the United States and Canada (whose participation had been reluctantly accepted by the USSR). The Helsinki text included promises that all signatories would promote freedom of movement and contact between their countries. The Soviet bloc's rulers, who had been very unwilling to make these promises, broke them brazenly, and persecuted the 'Helsinki groups' that asked them to comply. But, at the CSCE review conferences that continued until 1989, most western governments pressed for compliance, and the Helsinki message was spread among the east European peoples.

The 'iron curtain' was fraying. In 1989 it gave way. Unrest had become widespread in

Soviet control, 1945–89

Other communist states

0   miles   500

0   km   500

FINLAND
Helsinki

SWEDEN

IRELAND

◆Moscow

U S S R

Berlin
GERMANY   POLAND

CZECHOSLOVAKIA

SWITZ   AUSTRIA

HUNGARY

ROMANIA

YUGOSLAVIA

BULGARIA

ALBANIA

eastern Europe. The new Soviet leaders were in difficulty with their own crumbling economy; their army was disheartened by ten years of failure in Afghanistan *(50)*; they did not try
to reimpose hard-line regimes in eastern Europe by sheer force, as their predecessors had
done in 1956 and 1968. By the end of 1990, Poland, Hungary and Czechoslovakia had
elected non-communist governments; in Bulgaria and Romania the old communist bosses
had been removed; the East German state had been wiped off the map *(14)*. In 1991 the
Soviet Union itself fell apart, and even in Russia the communist party lost control *(16, 17)*.

Now that Europe was no longer forcibly divided, old links were gradually restored,
especially between the central European countries; but several new problems were posed.
Former Soviet satellites sought admission to their western neighbours' European Union and
to NATO, the North Atlantic alliance *(11, 12)*, but Russia strongly opposed enlargement of
the alliance. The violent disintegration of Yugoslavia *(15)* caused concern throughout
Europe – and showed that it was still difficult for Europe's governments to take united
action.

The CSCE machinery was developed in 1994 into the Organization for Security and Co-operation in Europe (OSCE), the membership rising to fifty-three with the admission of the new states that had emerged from the former USSR. Missions and monitoring teams from the OSCE were sent to some areas of conflict in former Yugoslav and Soviet territories; but Russian suggestions that the OSCE could take over the whole task of providing security in Europe, and that NATO could then be disbanded, found no favour with the western allies.

11 Atlantic Alliance

By the 1990s the North Atlantic alliance had surpassed, in size and duration, every other peacetime alliance in history. It had been formed as a counterweight to Soviet military power in Europe; and its future seemed uncertain now that eastern Europe was no longer under Soviet control and western Europe no longer feared Soviet domination. Yet new conflicts in Europe led to new calls for its services. The first combat operations ever carried out under its formal control were air strikes in 1994–5 against Bosnian Serbs, followed by management of the 60,000-strong peacekeeping force in Bosnia *(15)*. Between 1999 and 2004, the North Atlantic Treaty Organization (NATO) contributed to similar efforts in Kosovo, Macedonia and Afghanistan. After the 2001 terror attacks in the USA, NATO for the first time invoked article 5 of its founding treaty (an attack on one member is an attack on all), obligating all members to aid in actions against the terrorists.

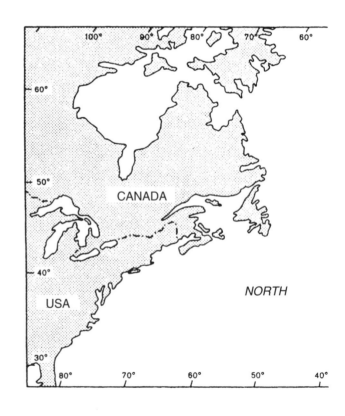

The alliance was first formalized in 1948, when Britain, France, Belgium, Holland and Luxembourg signed the Brussels Treaty. They saw that their five-member alliance could not adequately counterbalance the massive military strength that the USSR was maintaining in central Europe and they sought American support. Traditional American suspicion of 'entangling alliances' was overcome after several menacing Soviet moves – the Berlin blockade, the coup in Czechoslovakia, threats to Norway and Finland *(13, 14, 21)* . In 1949 the United States and Canada, the Brussels Treaty five, and Denmark, Iceland, Italy, Norway and Portugal concluded the North Atlantic Treaty. The 12 allies pledged joint resistance to any attack, in Europe, on any one of them. In 1950 the invasion of South Korea *(59)* increased their fears and led them to develop the more elaborate NATO.

Greece and Turkey joined the alliance in 1952; West Germany joined in 1955; Spain became the sixteenth member in 1982. (There was a further territorial extension in 1990, when the USSR was induced to agree that a united Germany could inherit West Germany's membership.) In 1966, France, resentful of the Americans' leading role among the allies, withdrew its forces from NATO's command structure – although it did not withdraw from the alliance – and the Allied headquarters were moved from France to Belgium. By 1996, however, the French had rejoined some of the NATO bodies they had quit in 1966. This reconciliation was largely due to the development of co-operative relations between the alliance and the Western European Union (WEU).

The WEU had been formed in 1955 by the five Brussels Treaty allies, Italy and West Germany; Greece, Portugal and Spain joined them later. One of the WEU's original purposes was to reassure both Germans and their neighbours that post–1945 rearmament would not foster a new German nationalist militarism. Many of its functions were later

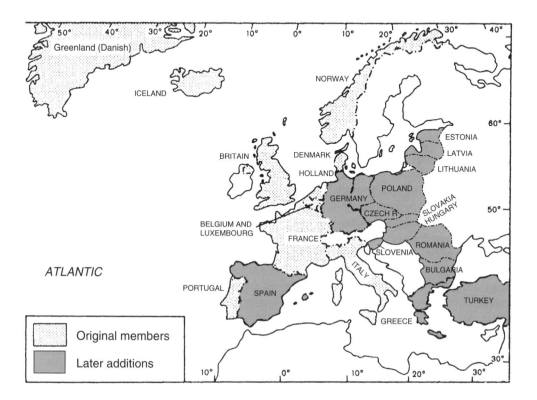

absorbed into the work of NATO; but in the late 1980s the ten-member grouping began to revive. Under its auspices, minesweepers were sent to the Gulf in 1988, other warships in 1990 *(48)*; from 1992 to 1996 it played a modest part in limiting violence in and around Yugoslavia. In 1993 its headquarters moved from London to Brussels, to facilitate co-operation with NATO. The revival of the WEU helped to satisfy French and other demands for a 'European defence identity' without alienating the United States – which, in fact, supported the idea that its European allies should, whenever they could, tackle European problems without seeking American help.

Only between Greece and Turkey *(26)* were there disputes among NATO members that ever brought them to the brink of war. But many quarrels about trade, fishing rights *(5, 22)* and other matters arose – as might be expected, in so large a grouping and over so long a period; and now there is no longer the fear of Soviet ambitions to provide an incentive to settle disputes quickly and restore allied unity.

After the collapse of communist power in Europe, the Atlantic allies faced one particularly tricky new problem. East European states were soon asking for admission to the alliance; but Russia gave warnings that it would react against an eastward expansion of NATO. In 1994 the alliance offered a form of association, 'Partnership for Peace', to east Europeans, to states that had emerged from the USSR in both Europe and Asia, and to Europe's 'neutrals' *(10)*. By 2005, 30 states had become 'Partners for Peace': five of the 'neutrals', 10 east European states, and all 15 ex-Soviet ones – including Russia. In 1997, NATO and Russia signed a separate agreement to co-operate on arms control, the resolution of conflicts in the Balkans, and other defence issues. This paved the way for east Europeans to move from Partner status to full membership of NATO. Poland, the Czech Republic and Hungary joined in 1999, followed five years later by Bulgaria, Romania, Slovakia, Slovenia and the three former Soviet Baltic states. The 'iron curtain' was no more.

12 European Unities

Problems that troubled European governments in 2005 included those raised by Turkish and Balkan requests for admission to the European Union, the integration of new and potential East European members, international migration *(24)*, and the failure of the proposed EU constitution. Heated debate on these questions sometimes obscured the amount of progress already made towards European unity.

Three things had motivated that progress. After the 1939–45 war, there was a strong desire to knit European states together in such a way as to prevent a recurrence of wars between them. There was an awareness of the need to band together against the Soviet power that had engulfed eastern Europe. And Europeans came to see that, if their relatively small domestic markets and economies remained separate, they would be at a great disadvantage in competing with the Americans and the Japanese.

Habits of working together began to develop in 1947, when the Organization for European Economic Cooperation (OEEC) was formed to handle the European Recovery Programme (or Marshall Plan) backed by American aid; it was succeeded in 1961 by the Organization for Economic Cooperation and Development (OECD), a group with more members and wider aims *(2)*. In 1948, Belgium, Britain, France, Holland and Luxembourg signed the Brussels Treaty; in 1949, with other Europeans, Canada and the United States, they signed the North Atlantic Treaty; in 1955 the Brussels Treaty five, West Germany and Italy created the Western European Union (WEU) – later joined by Greece, Portugal and Spain *(11)*. From 1949 on, the democracies sent ministers and members of parliament to the Council of Europe in Strasbourg (by 2005 it had 46 members, including all European states except Belarus). The conventions it drafted included one on human rights; to adjudicate on this, the European Court of Human Rights was created.

In 1948, Belgium, Holland and Luxembourg created a customs union, called 'Benelux'. During the 1950s the Benelux trio, France, West Germany and Italy, formed the European Coal and Steel Community (ECSC) and an atomic energy community, 'Euratom'. By the 1957 Treaty of Rome, they created the European Economic Community (EEC) – the 'Common Market'. Trade among the 'Six' was duty-free by 1968.

The Six agreed to merge the EEC, the ECSC and Euratom into a European Community (EC), and in 1967 they replaced the three executive bodies by a European Commission, based in Brussels. The Community shaped a common agricultural policy (CAP), promoted integration in other ways, and established a European Parliament at Strasbourg, whose powers were gradually enlarged, although final authority remained with the Council of Ministers, representing the member governments.

Britain and other countries which did not want to go as far or as fast as the Six formed the European Free Trade Association in 1960. By 1967 they had abolished tariffs on

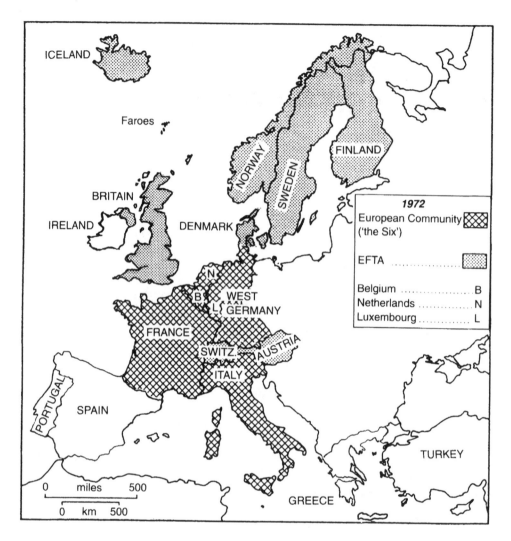

ICELAND

Faroes

FINLAND

NORWAY

SWEDEN

BRITAIN

IRELAND

DENMARK

1972
European Community
('the Six')

EFTA ....

Belgium .... B
Netherlands .... N
Luxembourg L

N

B

WEST
GERMANY

L

FRANCE

SWITZ

AUSTRIA

ITALY

PORTUGAL

SPAIN

TURKEY

GREECE

0   miles   500

0   km   500

non-farm trade between member countries. In 1973, Britain and Denmark left EFTA and joined the EC, and Ireland also joined. Greece joined the EC in 1981 and Portugal and Spain in 1986, giving it a membership of 12. The remaining EFTA countries – Austria, Finland, Iceland, Norway, Sweden and Switzerland – made free-trade agreements with the EC in 1972–3. As a result, duty-free trade in industrial goods was in effect by 1984 between all 18 members of the two groupings.

In a series of conventions signed at Yaoundé in Cameroon, at Arusha in Tanzania, and at Lomé in Togo, EC aid and access to the EC market were granted to 70 African, Caribbean and Pacific states. But this was offset by the protectionist CAP, and by the massive CAP subsidizing of European farming, which led to surpluses being 'dumped' outside the Community, doing grave harm to farmers in other regions (2). As a large component of an EU budget that now must cover infrastructure improvement in the new East European members, and a point of contention with the World Trade Organization, agricultural subsidies seem likely to decline in the future despite the current slow pace of reform.

In 1993 the 'European Union' emerged from the ambitious attempt to redesign the EC that had been made in the 1991 Maastricht treaty; and a 'single market', with full freedom of movement for people, capital and services as well as goods, was achieved – in principle. In 1995 the EU's membership rose to 15 when Austria, Finland and Sweden were admitted.

Iceland, Norway and Liechtenstein *(23)* became closely linked with the Union as members of the European Economic Area, established in 1994. Switzerland stayed out of both the EU and the EEA, but retained the free-trade links with the EU it had forged as a member of EFTA. There was now a 19-member free-trade zone with a population of 380 million.

After the fall of communism, the EU signed 'Europe agreements' with seven east European states and the three ex-Soviet Baltic states. Until 2004 the only ex-communist area that had gained entry was the former East Germany (whose absorption into a united Germany made the German position in the Union even more dominant than before [*13*]). In that year, eight eastern countries – the Czech Republic, Hungary, Poland, Slovakia, Slovenia and the Baltics – joined the EU, along with Cyprus and Malta. Romania and Bulgaria joined in 2007; Croatia, Macedonia and other Balkan countries are moving through the application process.

According to the 'Copenhagen criteria' of 1993, a potential member country of the EU must be a democracy with a market economy and respect for human rights, including the rights of minorities. EU members must also have agreed borders with their neighbours and be European, though what constitutes European has never been formally defined. All Balkan countries and several of the former Soviet countries are expected to seek admission eventually – but Morocco has also sought to join (though it is not a democracy), and Israel (a democracy, but without agreed borders) has considered applying. By 1996 the EU had formed a customs union with Turkey, and in 2005 Turkey (largely located in Asia) was approved as a candidate country. Discussion of the merits of admitting a secular Muslim country that would become the EU's most populous member, with borders extending to the Middle East, raised doubts as to whether Turkey will ever actually be admitted, and if so, on what terms.

In principle, the EU has aimed at enlargement. In practice, the big net contributors to its huge budget – Austria, Britain, France, Germany, Holland, Italy and Sweden – did not relish the prospect of new, low-income members with poorly managed economies lining up, with high hopes of generous handouts, alongside the existing group of big beneficiaries – Greece, Ireland, Portugal and Spain.

Inside the EU, strains showed between enthusiasts for integration, stubborn defenders of national sovereignty, and those who, while willing to edge towards closer union, saw danger in moving too fast. In some cases, the more cautious member states negotiated 'opt-outs'. In others, groups went ahead on their own. In 1995, France, Germany, Italy, Portugal, Spain and the Benelux three, under an agreement signed at Schengen in Luxembourg ten years earlier, ended (with some exceptions) the checking of passports at frontiers between them – thus creating 'Schengenland'. Since then, they have been joined by the other west European EU members except Britain and Ireland. Norway and Iceland (not EU members) are also in Schengenland, and all ten of the new EU members plus Switzerland have signed the treaty, providing for passport-free movement across most of the European continent by 2007.

A proposed constitution would have 'deepened' (further integrated) the EU by reducing the number of areas in which each member could veto policy changes without support from others, giving the EU more control over immigration policy, and establishing more clearly the primacy of EU law over national law. It was rejected by referendums in France and the Netherlands in 2005. Supporters of 'ever deeper union' continue to find little common ground with those opposing any encroachment on national sovereignty. No member has yet withdrawn from the EU, with the exception of the Danish territory of Greenland in 1985, which thereby regained control over its fisheries. (Several overseas provinces or territories of France, Spain and Portugal are part of the EU, while territories of other members have varying levels of ties with the Union.)

The EMU (Economic and Monetary Union) project, aiming at the adoption of a single currency, the 'euro', caused rifts not only between the warier members (notably Britain) and the enthusiasts, but also among the latter. Germany wanted a 'strong' euro; France, EMU's other chief sponsor, insisted on a 'weak' one; both eyed Italy and Spain nervously. It seemed that, to qualify for EMU, several governments would have to adopt unpopular economic policies, or questionable financial devices, or both. But by 1998 the exchange rates of euro-zone countries were fixed in preparation for the single currency. In 2002, the euro replaced national currencies in Austria, Belgium, Finland, France, Germany, Greece, Ireland, Italy, Luxembourg, the Netherlands, Portugal and Spain. It also circulates in several micro-states and parts of the former Yugoslavia.

In 2006 the economy of the EU was roughly equal in size to that of the US, but its population was half again as large. Unlike the US, most EU countries had birth rates well below the 'replacement level' – implying future population decline if immigration is not drastically increased.

13 Germany

In 1990, East Germany was united with a West Germany that had already become the biggest economic power in Europe. In area, however, the new united Germany was smaller than the Germany of 1937. The Nazi regime that held power from 1933 to 1945 had led Germany into a series of annexations and wars of conquest which, in the end, left it smaller, divided, and with hardly any contact between the two parts.

In 1938, Nazi Germany occupied Austria and, after a four-power conference in Munich, forced the cession of Czechoslovakia's Sudeten region, running along its frontier with Germany. In 1939 it seized the rest of Czechoslovakia; demanded Danzig (now Gdansk), which was then a 'free city'; and, with Danzig as a pretext, invaded Poland – after making a pact with the Soviet Union (the 'Stalin–Hitler pact') under which the two powers shared out Poland. Britain and France had given guarantees to Poland; so they declared war on Germany. At one stage in the 1939–45 war Germany, with its allies, controlled most of Europe; its forces had conquered an area that stretched from Norway to Greece, and from France to the Volga, deep inside Russia. But by 1945 the Nazi empire was destroyed, and western and Soviet armies occupied Germany.

Austria was restored to independence, although American, British, French and Soviet troops remained there until 1955. The Sudetenland was returned to Czechoslovakia, and its German inhabitants were expelled. France occupied the Saar, but returned it to Germany in 1957. The Soviet Union annexed northern East Prussia and, expelling the inhabitants of the other parts of Germany east of the line of the Oder and Neisse rivers, assigned these areas to Poland.

The Germany that remained was divided into American, British, French and Soviet zones; Berlin was similarly divided into four sectors. In the Soviet zone, communist rule was imposed. When the western allies, unable to obtain Soviet co-operation, introduced reforms in their zones in 1948, a Soviet blockade of West Berlin forced the Allies to mount a year-long airlift to feed its 2 million people. Soviet obstruction of the western routes to Berlin was often used as a form of pressure later, most notably in 1961.

In 1949 the western zones were united as the Federal Republic of Germany, with its capital at Bonn. The Soviet zone's communist regime was styled the German Democratic Republic. In the 1950s some 6 million people escaped from east to west; to stop this flight, the communists built a massive 30-mile wall across the middle of Berlin in 1961; they also cleared a mined and guarded 'death strip' all along the border between East and West Germany.

A 400,000-strong Soviet army remained in East Germany, which was made a Soviet 'ally', first in a bilateral treaty and later in the Warsaw Pact. West Germany joined NATO in 1955, and, except in Berlin, the western forces' occupation status was then ended.

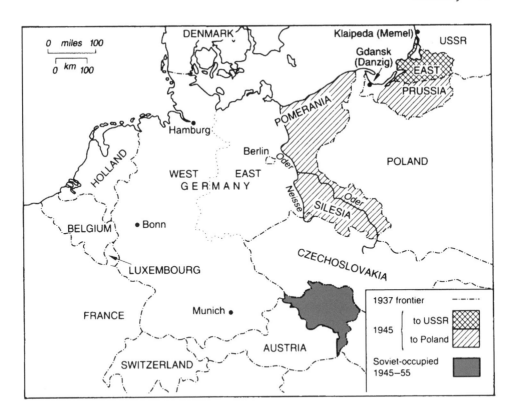

In the 1970s, West Germany made treaties with the USSR, Poland and East Germany; but the 17 million East Germans were still denied free contact with the 62 million West Germans. In the later 1980s, reforms in some east European countries permitted 700,000 of their ethnic Germans to get out to the west; East Germans demanded similar permission, and flocked to West German embassies in eastern Europe. In 1989, when Hungary opened its frontier with Austria, 200,000 of the East Germans got out that way in a few weeks. Their demoralized rulers, who were no longer sure of Soviet backing for repression, opened the Berlin wall; but the outflow went on until they promised a free East German election. The voting, in 1990, produced a non-communist government; by the end of that year German unification had been negotiated and accomplished.

In American–British–French–Soviet–German negotiations, it was agreed that the new unified Germany would confirm its acceptance of the Oder–Neisse frontier and its renunciation of nuclear weapons; that the Soviet forces would leave by 1994, Germany paying $6 billion towards the cost of removing them; and that West Germany's membership of the NATO alliance and the European Union would be inherited by the new Germany. The Soviet Union had long demanded that any united Germany must be 'neutralized'; but now the East European states assured it that a united Germany in NATO would be a more comfortable neighbour than one that might swing towards nationalistic policies if it had no allies to support and restrain it.

Communist rule had left eastern Germany in a sorry state, and since 1990 the western part of the country had been obliged to pay more than $1 trillion in subsidies to modernize the eastern part. This burden on the German economy also had indirect effects on other

European countries and led to the realization that the other potential post-Cold War unification project (Korea) would be more difficult than previously assumed. But, in general, Germany's European Union partners were less concerned about the impact of its current difficulties than about the longer-term prospect of a unified German economy playing a strongly dominant role in the partnership. Germany has in recent years been the world's leading or second (after the US) exporter.

In 1991 it was decided to move the government back from Bonn to the former capital, Berlin; the transfer became official in 1999. In federal Germany the governments of the 16 states (*Länder*) have wide powers, and members of state governments form the federal upper house; the placing of the federal court in Karlsruhe and of the central bank in Frankfurt (also home to the European Central Bank) has been another sign of a wish to avoid excessive centralization.

14 Central and Eastern Europe

Near the end of the 1939–45 war, at a conference at Yalta in the Crimea, the Soviet Union agreed with the United States and Britain that the nations liberated from the Nazi German empire *(13)* should be enabled to create 'democratic institutions of their own choice'. However, it flouted the Yalta agreement. In the six countries that its army occupied, it imposed 'satellite' communist regimes (Czechoslovakia had a brief respite, but only until 1948). Elections were rigged; many democratic leaders were killed, many jailed; non-communist parties were purged and became docile 'fellow-travellers'. The Soviet army did not occupy Yugoslavia or Albania, but local communists installed similar regimes *(15)*.

In 1955, when Austria committed itself to neutrality *(13)*, the USSR urged Germany to do the same, but its propaganda about neutrality led to stirrings in eastern Europe, and in 1956 Hungary tried to become neutral. The Soviet army crushed this Hungarian revolution – at a moment when the world's attention was distracted by the Suez conflict *(43)*.

In the 1960s many communists in Czechoslovakia were repelled by their own regime's brutality. New leaders tried to give it 'a human face', assuring the USSR that they would not turn neutral. The Soviet rulers could not allow a satellite state to make even modest reforms. In 1968 their army moved in – dragging East German and other satellite troops with it – and imposed a new puppet government.

The 1970s brought new outbreaks of protest in Poland. A wave of strikes in 1980 forced the communists to let the workers form a free labour organization, Solidarity – whose leader, Lech Walesa, had headed its first activities in Gdansk (formerly Danzig). But a military coup soon silenced Solidarity's demands for reform. Its leaders were seized, strikers were forced back to work under threat of death, and the army chief, General Jaruzelski, became head of the communist party and the government.

Through all this, East Germany's and Bulgaria's governments toed the Soviet line. Romania's did not always do so, but its rule was equally repressive. After 1956 it clamped down on the 2 million Hungarians in Transylvania, abolishing that region's autonomy and deporting many of them to the Dobruja swamplands. New discriminatory measures were imposed on them in the 1980s, and thousands fled to Hungary.

Another victimized minority was the Muslims in Bulgaria. In 1972–3 Bulgaria's 300,000 Pomaks (Muslims of Slav origin) were made to renounce their Muslim names. In 1984–5 the forcible name-changing was extended to the 1.2 million Muslims of Turkish origin; mosques were closed, and the use of Turkish dress or language was banned. In both these drives, many people were killed. As late as 1989, 325,000 Muslims were driven from their homes and forced to flee to Turkey.

However, in 1989 the east European ice cracked. About 250,000 East Germans escaped through Hungary, and their rulers were obliged to open the Berlin wall *(13)*. Poland's rulers

'Independent' states effectively controlled by the Soviet Union until 1989

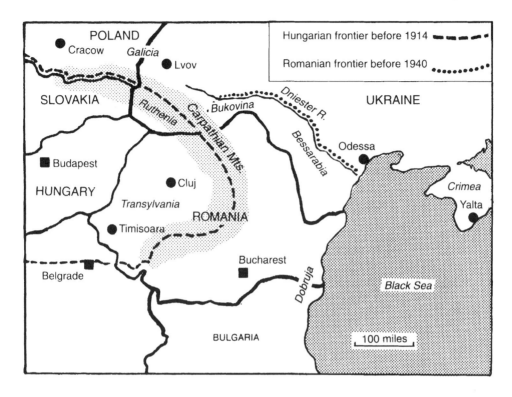

had to permit a semi-free election. Nearly all the seats filled by free voting were won by Solidarity, which formed a government (Jaruzelski remained president, but in 1990 Walesa was elected to succeed him). Romania's ruling Ceausescu family, thuggish to the end, went down in bloodshed in December 1989.

By the end of 1990 elections had been held, and non-communist governments had taken over, in Poland, Czechoslovakia, East Germany and Hungary. Elections had also been held in Bulgaria and Romania; but their communists, under new labels and new leaders, and promising to mend their ways, contrived to retain power.

The Romanians were not able to vote their old-guard president and government out of office until 1996; Bulgaria's ex-communists were ousted for a time, but returned to office for a period in the mid-1990s. By then ex-communists had made comebacks in elections in Poland and Hungary. But these were more truly reformed than the Balkan ex-communists; they respected civil liberties, encouraged the private sector and sought links with western countries.

As early as 1989 one east–west link, intended to revive regional co-operation in central Europe, had been created: the 'Pentagonal' (five-member) grouping of Austria, Czechoslovakia, Hungary, Italy and Yugoslavia. It became 'Hexagonal' when Poland joined two years later, was renamed the Central European Initiative, and had expanded to eighteen members by 2006. In 1991 the last Soviet troops withdrew from Czechoslovakia and Hungary (in Poland, some Russian forces remained until 1993); and a conference at Visegrad, near Budapest, created a grouping of Czechoslovakia, Hungary and Poland, all of them seeking admission to the EU and NATO. By 2004, eight countries in the region had joined the EU *(12)* and ten had been admitted to NATO *(11)*.

Czechoslovakia was divided, peacefully, in 1993, tensions having increased between the Slovaks, among whom conservative and communist elements were both strong, and the westward-oriented Czechs, who now set up the Czech Republic. After this 'divorce', Slovakia's relations with Hungary deteriorated, and in 1995 Slovakia adopted new language laws, to the disadvantage of its 575,000 Hungarians.

Economic growth has followed governmental reform in the region since 1990. Lower wages, proximity to western European markets, EU funding for westward transport links, and the promise of EU membership (realized in 2004) all encouraged foreign investment. Cheaper air fares and fewer customs barriers have boosted tourism and industry alike, from the historic Prague to the car factories of Slovakia.

Democratic reform and the desire to join the EU and NATO encouraged better treatment of minorities and greater openness about past injustices in the later 1990s. Both Slovakia and Bulgaria liberalized their language laws, and Bulgaria apologized for previous action against its Turkish minority. Efforts began to return, or provide compensation for, property confiscated from Jewish owners by the Nazis or national governments during the Second World War.

The status of the 5–10 million Roma ('gypsies') remained far from ideal. Discrimination in hiring and education and geographic segregation have ensured exceptionally high rates of illiteracy, poverty and unemployment for Europe's largest non-immigrant minority. Governments in the region have only recently begun to address these problems, aided by the EU and international organizations.

After the collapse of the Habsburgs' Austro-Hungarian empire in 1918, several countries had been left with Hungarian minorities. In Serbia's Vojvodina region, Hungarians formed a quarter of the population. The abolition of the region's autonomy in 1989 and the wars in the former Yugoslavia made their position difficult; many fled. However, the 200,000 Hungarians in Ukraine – mostly in its Transcarpathian region *(18)* – were granted cultural autonomy; and in 1996 Romania signed a treaty with Hungary in which it promised to respect its Hungarians' rights.

15 Former Yugoslavia, Albania

The wars that began in 1991, when Yugoslavia broke up and five new sovereign states emerged, were Europe's most violent conflicts since 1945. In 2006, although foreign intervention and mediation had stopped the fighting, the presence of international peace-keeping forces was still found necessary to prevent further large-scale bloodshed.

After the 1939–45 war, in which Yugoslavia and Albania came under German and Italian control, communists took power there. They were not occupied by Soviet forces, but at first they followed the Soviet line. (Yugoslavia was then quarrelling with the western allies over Trieste, a mainly Italian-peopled city; British and American forces stayed there until 1954, when Yugoslavia agreed to let Italy keep Trieste.)

In 1948, Yugoslavia broke with the USSR. This rift disrupted the Albanian–Yugoslav–Bulgarian backing of communist rebels in northern Greece, the last of whom withdrew into Bulgaria in 1949. With western help, Yugoslavia withstood Soviet pressure for several years. It cultivated links with 'third-world' countries, and in 1961 the 'non-aligned' movement was founded at a conference in Belgrade *(6)*.

Albania sided with the USSR against Yugoslavia in 1948, then with China against the USSR in 1961. In 1978 it broke with China, and for 12 years it was the most isolated of communist states, still ruled by 'Stalinists' – who, however, failed to end old internal antagonisms: between Albania's Gheg northern highlanders and Tosk southerners, between its Muslim majority and Christian minority. And the presence of a Greek minority in southern Albania complicated its relations with Greece.

Meanwhile Yugoslavia's 2.5 million Albanians posed problems. Serbia, one of Yugoslavia's six republics, contained an 'autonomous region', Kosovo, whose people were mainly Albanian. In 1989, after recurrent violence, Serbia's communist rulers abolished Kosovo's autonomy. Discontent and a wish to break away increased among its Albanians.

Then the 'wind of change' that was blowing through eastern Europe reached Yugoslavia. In 1990 its Slovenian and Croatian republics held their first free elections and ousted their communist bosses. Their demands for a loosening of the federation were rejected; Serbia, the largest of Yugoslavia's republics, was still run by communists, and Serbs, although only 35% of Yugoslavia's population, dominated its armed forces. In 1991, Slovenia and Croatia declared their independence.

The Serb-led Yugoslav army withdrew from Slovenia after making only a brief attempt to subdue it. Independent Slovenia was then left in peace. In Croatia, however, the Yugoslav army helped the Serbs, who made up 12% of the population, to take over areas forming a third of its territory (expelling all Croats from those areas). A 'republic of Serbian Krajina' was proclaimed, with its capital at Knin. In 1992 the United Nations secured a

ceasefire in Croatia and installed a small UN force. Serb aspirations to create a 'greater Serbia' then turned to Bosnia.

In Bosnia (formally, Bosnia and Hercegovina) the 4.5 million people were 31% Serb, 17% Croat and 44% Bosniak (Muslim). They all spoke Serbo-Croat and were of common origin. The only clear division was religious: the Croats were Catholic, the Serbs Orthodox; the Muslims' ancestors had been converted to Islam during 400 years of Turkish rule. Bosnia's first free election, in 1990, produced a government in which the three communities were proportionately represented. But in 1992, when Bosnia's Muslims and Croats voted for independence, its Serbs set up a rival government and, with help from Serbia, took over large areas, killing or expelling their non-Serb inhabitants. By mid-1993 they held 70% of Bosnia's territory; and they had been besieging its capital, Sarajevo, for a full year. Meanwhile Bosnia's Croats, with help from Croatia, had seized areas – mainly in Hercegovina, the southern region – in which they proclaimed a 'Croat republic of Herceg-Bosna'.

A small UN force protected food-aid convoys which helped Sarajevo to hold out and

saved about 3 million people from starvation, but mediation by the European Union and the UN failed to restore a united Bosnia. By 1994, however, UN economic sanctions imposed on Serbia had led it to reduce its support for the Bosnian Serb forces; and the Bosnian Croats and Muslims had been induced to stop fighting each other. In 1995 the Bosnian Serbs' massacre of 8,000 Muslims at Srebrenica outraged world opinion, and their persistent attacks on Sarajevo brought retaliation by NATO aircraft which made them pull back. The Muslims and Croats then made gains at their expense in western Bosnia. Meanwhile Croatia had recaptured all the Serb-held parts of its territory except Eastern Slavonia, where another UN force was installed.

In 1995 negotiations at Dayton in the United States produced a peace pact for Bosnia. In principle it would be one country; in practice it would comprise two 'entities'. The Serbs, who then held 48% of its territory, were assigned 49%, the rest going to the Muslim–Croat 'federation'. Peacekeeping forces were organized under NATO direction. Elections in 1996 produced a new Bosnian presidency and parliament; but the three communities still controlled their separate areas and the UN remained in ultimate control, charged with oversight of the procedures agreed at Dayton. Brutal 'ethnic cleansing' – not all by Serbs – and postwar migration caused almost complete segregation between the two 'entities'.

In 1991, Macedonia followed Slovenia and Croatia in declaring its independence. A small UN force was sent to Macedonia to discourage 'Yugoslav' (i.e. Serb) moves against it, and by 1996 these neighbours had established normal relations. Greece, however, accused the new state of aiming to annex the Greek region of Macedonia, and demanded that it should change its name and its flag (which showed an ancient Macedonian symbol). Landlocked Macedonia, whose main outlet is Greece's port of Salonika, was hard hit by a Greek ban on trade, which American mediation ended in 1995. Macedonia agreed to change its flag, but the dispute about the name was unresolved. The new state had been admitted to the UN in 1993 under the name of 'Former Yugoslav Republic of Macedonia' (FYROM). It faced yet another problem, in the shape of a large Albanian minority, concentrated in areas near the Albanian frontier, where demands for autonomy led to rebellion in 2001. NATO intervention ended the conflict after a few months, and a peace agreement included some devolution of power to local governments and official status for the Albanian language. In 2005 the EU accepted Macedonia (officially FYROM) as a candidate for membership.

The remainder of Yugoslavia continued to use that name until 2003, when it was renamed Serbia and Montenegro. Montenegro voted for independence in a 2006 referendum. The southern region of Kosovo had less success in breaking away from Serbia. Though 90% Albanian in population today, Kosovo was the centre of the Serbian state in medieval times. It was part of Serbia during the communist period, gaining autonomy in 1974. After this was withdrawn by Serbia in 1989 (see above), the Albanians sought renewed autonomy or independence, first by peaceful means. Civil war in the late 1990s led to NATO air strikes in 1999 that forced the Serbian military out of Kosovo. Since then, Kosovo has been controlled by the UN while remaining formally part of Serbia. The success of future negotiation on independence or greater autonomy is linked to Serbia's desire for EU membership – not possible with an ongoing territorial dispute.

16 Former Soviet Union

The most dramatic event of the early 1990s was the breakup of the Union of Soviet Socialist Republics. It had been the world's largest country, with its third-largest population – an acknowledged 'superpower', with huge armed forces and a nuclear arsenal matched only by that of the United States. From 1945 to 1989 it had dominated half of Europe. In 1991 it vanished from the map – and was replaced by 15 sovereign states, one being Russia.

The USSR was never what its title asserted. It was under even tighter central control than its predecessor, the pre-1917 tsarist Russian empire. But on paper it was a federation of 15 'union republics', each, in theory, free to secede. The biggest, the Russian republic, was itself a federation, containing 16 'autonomous republics' and 15 'autonomous' regions and areas. Ostensibly, forms of self-government were thus provided for dozens of national groups, ranging from the Baltic Estonians to the Yakuts of eastern Siberia. In reality, the communists' political monopoly meant that power was concentrated in Moscow.

Long-suppressed desires for real national independence were quickened in 1989 by the sight of east Europeans emerging from Soviet domination. In 1990 the Soviet republics were allowed to hold the first elections in which the communists could be challenged. Wherever voting was really free, the communists lost. The republics promptly asserted their right to secede. In 1991 the USSR recognized the independence of the Baltic republics *(18)*; the other republics, including Russia, agreed to form a 'Commonwealth of Independent States'; and on 25 December the Soviet Union was voted out of existence by its own parliament.

Of the USSR's population, nearly half were Russians. A fifth were Ukrainians and Belorussians – Slavs of much the same stock as Russians, but with distinctive languages and traditions, partly resulting from long association with Poland and Lithuania *(18)*. The rest were non-Slavs, and these included central Asian peoples whose culture is Muslim *(20, 54)*.

Tsarist Russia had been an exporter of grain, and in its later years it had Europe's fastest-growing economy. By the 1970s the USSR had to import more grain than any other country; by the 1980s its total output and living standards were falling. Much of the output of its industries was not 'goods' but 'bads' – products worth less than the materials used to make them. When Mikhail Gorbachev reached the top of the party hierarchy in 1985 he called for reforms, but the hierarchy obstructed them. Both the centralized economy's failure and the failure to reform it encouraged the separatism that eventually tore the Soviet Union apart.

Even the USSR's successes had a dark side. It became the world's biggest producer of oil and second-biggest producer of gold, mainly by developing its Siberian and Asian

resources. That development involved labour from the 'Gulag Archipelago', the array of prison camps ranging from Karaganda and Vorkuta to the Kolyma. In Stalin's time 18 million people died in the *gulags*. In the 1980s, *gulag* labour helped to build the BAM railway *(54)* and the pipelines carrying Siberian gas to Europe *(3)*.

After 1991, some of the agreements that Russia made with other ex-Soviet states

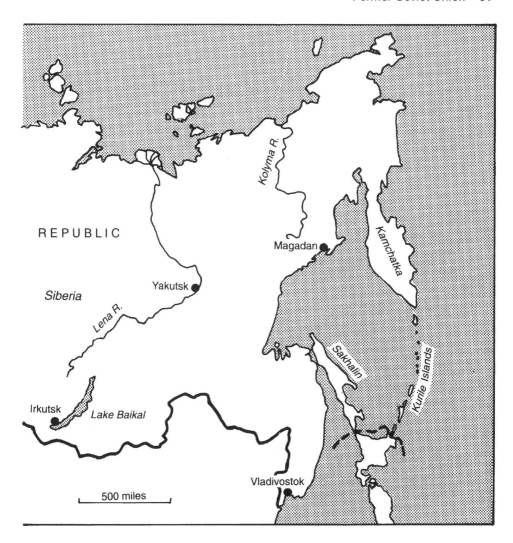

REPUBLIC

Siberia

Yakutsk

Lena R.

Kolyma R.

Magadan

Kamchatka

Sakhalin

Kurile Islands

Irkutsk *Lake Baikal*

Vladivostok

500 miles

suggested a wish to reimpose its control on what it called 'the near abroad'. In 1996 its communists (still a strong party) even called for the restoration of the USSR. That call went unanswered. At times, however, Russia put heavy pressure on republics that needed its oil; and, when it sent troops to help governments in 'the near abroad', some of them looked like staying there for a long time *(19, 20)*.

17 Russia

Before 1991, Moscow was the capital of the Soviet Union, which had nearly 300 million inhabitants of whom 147 million (49%) were Russians. It was also – and after 1991 it still was – the capital of Russia, a country with almost 150 million people of whom 122 million (81%) were Russians. Russia's population peaked soon after the collapse of the Soviet Union and since then has been declining, the result of sharply lower life expectancy among Russian men, a low birth rate, and emigration. The negative demographic effects of poverty, the end of Soviet restrictions on alcohol, reduced access to basic health care, and AIDS are not balanced by continued immigration of ethnic Russians and others from the ex-Soviet 'near abroad'.

Outside Russia, as of 1991, there were 25 million Russians in the 14 other ex-Soviet

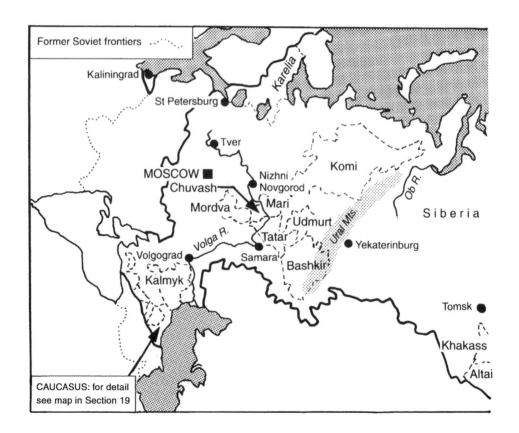

countries; among them, 11 million in Ukraine and 10 million in central Asia. The Russian government's concern for them (understandable in some cases, suspect in others) strongly influenced its relations with the fourteen other countries. In 1996 there were also more than 100,000 Russian soldiers in those countries, mostly stationed near the former Soviet frontiers, in Belarus, Moldova, the Caucasus and central Asia, and on the Crimean peninsula in Ukraine. Some of the explanations offered for their presence were plausible, but Russia's apparent eagerness to take over the Soviet Union's former responsibilities worried many people *(18–20)*. A decade later Russia had withdrawn from many of its bases in the former Soviet countries and closed others in Vietnam and Cuba – leaving Syria as its only outpost outside the ex-Soviet region – but had opened new outposts in Tajikistan and Kyrgyzstan (where there is also an American base).

Inside Russia, after 1991, its government faced complex problems in regard to the many national groups in its non-Russian population (which included 18 million Muslims). The 16 former 'autonomous republics' of Soviet Russia *(16)* were allowed to become fully acknowledged (but not independent) 'republics', and five other minority areas achieved a similar status. In a series of agreements and measures, including the federation treaty of 1992 and the new constitution of 1993, wider powers were granted both to these 21 republics and to Russia's regional authorities.

Some national groups remained dissatisfied. Chechnya demanded complete independence, and in 1994–6 the Chechens fought the Russians to a standstill. Fears that some of Russia's other republics might follow this example proved valid as conflict spread through the Caucasus – often with Chechen involvement *(19)*. Most republics were small, or poor,

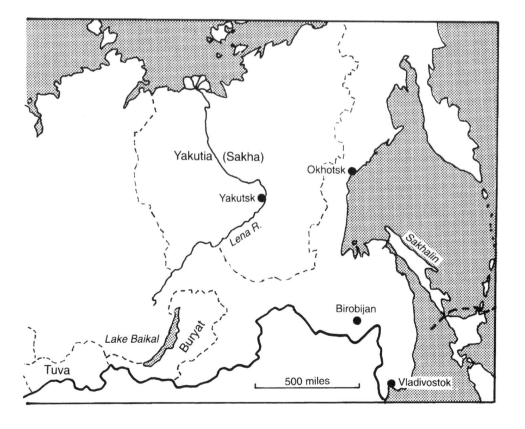

or both (but being small and poor had not stopped Chechnya from fighting a war that made it even poorer). In nine of the republics, Russians outnumbered the people whose name the republic bore; Buryatia's population was 70% Russian, Karelia's 74% Russian.

Yakutia (Sakha), although neither small nor poor – it mines gold and produces 99% of Russia's diamonds – had only a million inhabitants, half of them Russians, and seemed content to obtain a bigger share of the profits from diamond sales. Bashkortostan and Tatarstan (each with a population of 4 million, about 40% Russian) demanded extra constitutional rights, and had enough leverage to get them; these republics in European Russia between the Volga and the Urals, homes to the Muslim Tatars and Bashkirs, had plenty of oil and gas, and they straddled Moscow's lines of communication with Siberia. Bashkortostan signed the 1992 treaty only after it had secretly negotiated a special deal. Tatarstan held back until 1994, when it signed a bilateral treaty with Russia after long haggling.

One of the 'national' entities marked out on Russia's map in the Soviet era never achieved authenticity. A 'Jewish autonomous region' was created in 1934 – distant from any large Jewish community – in a bleak area in Siberia's Far Eastern region, around Birobijan on the Chinese frontier. In 1991 only about 10% of Birobijan's population of 220,000 was Jewish; and by 2006 only a quarter of the Jews remained. There were still about a million Jews in the ex-Soviet countries, although more than a million have left since 1989, when the Soviet government began to permit their emigration.

A unique feature of Russia's Soviet-era maps was the concealment of 35 'secret cities' with a total population of 2 million. New maps may show some of them: e.g. Seversk, north of Tomsk, and Lesnoy, near Yekaterinburg. From 1942 onwards they had been built, inside heavily guarded and electrified perimeter fences, for development and production of nuclear, chemical and biological weapons. The ban on any mention of their existence meant that many releases of radioactivity, anthrax leaks and other environmental disasters were hushed up.

On the other hand, many names that did appear on the Soviet-era maps have disappeared. After 1991 the Russians wanted to recover their lost history. One of the ways their rulers had tried to wipe it out was by giving old cities the names of approved communist personalities. Now Leningrad has reverted to St Petersburg, Kalinin to Tver, Gorky to Nizhni Novgorod, Sverdlovsk to Yekaterinburg, Kuybyshev to Samara. Smaller places that had been renamed to honour Andropov, Brezhnev, Chernenko, Gottwald, Zhdanov and others have also retrieved their real names. (The countless places called Stalin, Stalino, Stalinsk, Stalingrad, Stalinogorsk, etc., had all been 'de-Stalinized' when the communists' 'party line' was changed in the 1960s.)

18 Baltic to Black Sea

Estonia, Latvia and *Lithuania*, which had won independence when the old Russian empire foundered in 1917, were annexed in 1940 by the USSR. In 1990 they reasserted their independence, declaring that the annexations had been illegal. The Soviet Union, after brief attempts at suppression, recognized the three Baltic states' independence in 1991; but much negotiation was needed to bring about the withdrawal of Russian forces, which was not completed until 1995. Problems also arose over the Russians who by 1991 had come to form 31% of Estonia's population and 36% of Latvia's (Lithuania had a Russian minority of only 8%). Angry gestures from Moscow followed Estonia's decision that post-1940 Russian immigrants must pass a language test to become Estonian citizens. Many could not, and as a result more than 10% of Estonian residents are not citizens of any country.

One problem for the Lithuanians was that their country lay between Russia's *Kaliningrad* 'exclave' and its main territory. In 1945 the USSR had annexed half of Germany's exclave of East Prussia – around Königsberg, then renamed Kaliningrad – and designated it as part of Soviet Russia *(13)*. Its German inhabitants were expelled, and it became a military base area (with a naval base at Baltisk). In 1996 there were still 90,000 Russian soldiers there; the transit of Russian troop-trains across Lithuania led to disputes about frontier formalities; and the Poles reacted testily to the Russians' request for a 'corridor' through Poland as an alternative supply line for their forces in Kaliningrad. The EU sought tougher border controls when Poland and Lithuania joined in 2004, but Russia was able to negotiate a compromise allowing transit for its citizens.

In sharp contrast to the Baltic states, *Belarus* (formerly Belorussia or Byelorussia), after becoming independent in 1991, clung to Russia more tightly than any other ex-Soviet country. Only 13% of its 10 million people are Russians, but its language closely resembles Russian and cultural ties are strong. Belarus depended on Russia for oil and other essentials. It reunited its monetary system with Russia's, and joined Russia in a customs union. In 1996 it joined it in a (two-member) 'Commonwealth of Sovereign Republics'. In effect, Russia was to prop up Belarus's economy and to control its frontiers, armed forces and diplomacy. Unlike its ex-Soviet and east European neighbours, Belarus has not moved towards democracy.

Ukraine has shared a lot of history with Belarus. Over five centuries, both experienced Lithuanian, then Polish, and finally Russian rule; after 1921, Poland held their western regions until the USSR seized them in 1939; in 1941–4 both were devastated by the German invaders. But after 1991 they took different roads.

Of Ukraine's 52 million inhabitants, 22% were Russians; in some of its eastern areas, they made up 40% of the population. But, to the indignation of those Russians who wanted to see the former USSR's three Slav republics reunited, Ukraine proved less docile than

Belarus. In 2004 its 'orange revolution' forced a pro-Russian prime minister to resign after a rigged election.

Ukraine's inability to pay for the oil and gas it needed enabled the Russians to apply pressure, culminating in a cutoff of gas supplies in 2006 when Ukraine refused to pay a price closer to market value. Though the dispute was quickly resolved, it highlighted EU dependence on Russian energy supplies, because the same pipelines supply both Ukraine and western Europe.

Sharp disputes arose over the *Crimea*. In 1954 it had been transferred from Russian to Ukrainian control. Its Tatar inhabitants had been deported to central Asia in 1945; after Stalin's death many tried to return home, but found no welcome. In 1991, 75% of its population of 2.7 million was Russian; the 1990s saw the return of 200,000 Tatars from central Asia. Ukraine granted Crimea autonomy; but its parliament called for independence – which would really mean a return to rule by Russia. In particular, Russia wanted control of the Sevastopol naval base (the objective of the Franco-British forces in the Crimean War of 1853–6). The former Soviet Black Sea fleet lay there – rusting – while Russia contested Ukraine's demand for a share of it. This was gained in 1997, when Ukraine agreed to a twenty-year Russian lease of the naval base.

The Galicia region in western Ukraine, around Lvov, had been under Austrian rule before 1918, acquiring a 'central European' character; in 1990 its people were foremost in calling for Ukrainian independence. To its south, across the Carpathian range, is Ruthenia (Sub-Carpathian Ukraine, or Zakarpatskaya); having known Hungarian and Czechoslovak rule before the USSR annexed it in 1945, it has retained a distinct identity, and its people speak a Slovak-influenced form of Ukrainian.

In *Moldova* (Moldavia), which the USSR took from Romania in 1940, two-thirds of the 4.4 million inhabitants speak Romanian; 13% are Russian, 14% Ukrainian; 3% are Gagauz, of Turkish origin. After 1991, when, for a while, Moldova looked like joining Romania, civil war broke out. Russian troops came in and stopped the fighting, but then settled in, protecting a breakaway 'Transdniestria republic' along the Dniester's east bank. This sliver of territory is ruled communist fashion, brutally and incompetently, by Russians – who together with Ukrainians make up half the population – and until its status is resolved, Moldova cannot progress towards EU membership. The Gagauz, who had also rebelled, were pacified in 1995 by an offer of autonomy for their region in the south.

19 Caucasus

In the Caucasian regions, violent conflicts between national groups had begun before the Soviet Union broke up. The Muslims of *Azerbaijan* and the Christians of *Armenia* were virtually at war by 1990; the fighting had started in 1988 in Nagorno-Karabakh, a mainly Armenian-peopled hill area in Azerbaijan. Until 1991, Azerbaijan, still ruled by hard-line communists, had support from Moscow, but in 1992 the tide turned in the Armenians' favour. When international mediation helped to stop the war in 1994, they held nearly all of Nagorno-Karabakh and all the territory between it and Armenia. But 400,000 Armenians had had to flee from Azerbaijan, and Armenia had had to let in some Russian troops, who watched over its Turkish frontier. (The Turks, sympathizing with Turkic Azerbaijan although reluctant to get too involved, closed their frontier with Armenia during the war.)

Georgia, Stalin's homeland, where vestiges of his 'personality cult' had persisted after his death, was rather unexpectedly quick to demand independence when the USSR began to crumble. But a separatist revolt promptly broke out in Abkhazia. The Abkhazians, themselves a very small community, were helped by fellow Muslims from neighbouring areas and, more obliquely but more effectively, by the armed forces first of the USSR and later of Russia. In 1993, Georgia had to agree to a virtual Russian takeover of Abkhazia. Meanwhile there had been a civil war among the Georgians themselves and a revolt in South Ossetia, whose people wanted to join the Ossetians north of the Caucasus range. Here, too, Russian forces took a hand; under their protection, South Ossetia was separated from Georgia in 1992. Russia also acquired a naval base at Batumi, in Muslim Ajaria, and by 1996 there were 30,000 Russian soldiers in Georgia.

Independent Georgia was ruled by a former Soviet foreign minister until 2003, when a fraudulent parliamentary election triggered protests (the 'rose revolution') that compelled him to resign. New elections brought the pro-western opposition to power. Support for the new government extended into Ajaria, where popular pressure forced the local dictator out in 2004. Russia agreed to withdraw all of its troops from Georgia by 2008 – at which time Georgia intends to seek NATO membership.

While the conflicts south of the Caucasus mountains generally strengthened Russia's hand in dealing with its neighbours, it met a severe setback north of the range, within its own borders, in *Chechnya*. The million Chechens, the most numerous of the North Caucasian peoples, had a long record of resistance to Russian rule. In 1944 they, and other Muslim mountain peoples, had been deported to central Asia; they were not allowed to return until after 1957. The Chechens proclaimed an independent republic, 'Ichkeria', in 1991. They rejected Russia's offers of autonomy and withstood its pressures. (In 1993, Chechens were expelled from Moscow; however, it may be noted that their mafia-type gangs there had become notorious.) In 1994 the Russian army was sent in, but fierce

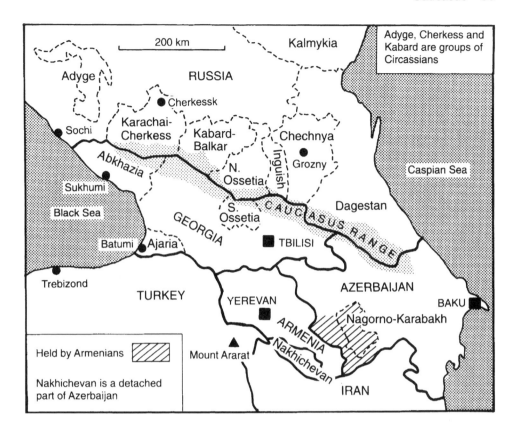

resistance inflicted on it casualties estimated as much heavier than those it had suffered in Afghanistan.

In 1996 a ceasefire was negotiated, on terms which some Russians thought humiliating, although many felt relief. By the end of the year all the Russian troops had been withdrawn and all of Chechnya was in the Chechens' hands. Under the ceasefire terms, its final status was to be decided five years later. However, following terrorist bombings in Moscow in 1999 and Chechen involvement in a rebellion in Dagestan, Chechnya's eastern neighbour, the Russian army returned. Conflict has also spread to Chechnya's western neighbours. Chechen involvement in rebellions, Muslim militant activity, and inter-regional wars have drawn strong Russian response to Ingushetia, North Ossetia – where a terrorist takeover of a school left more than 300 people dead in 2004 – and Kabardino-Balkaria.

The Ingush, the Chechens' Muslim neighbours on the west, had also been deported in 1944, and on their return they found some of their land occupied by the mainly Christian and more Russified Ossetians. In 1992 large-scale fighting broke out; Russian troops intervened in favour of the Ossetians, defeating the Ingush but not resolving the dispute. The Terek Cossacks, whose local centre was at Vladikavkaz in North Ossetia, also became involved. Settlements of Cossacks – no kin of the Kazakhs, although the names have a common Turkic root – were scattered across the North Caucasus. Their position was precarious; they claimed to be a people clearly distinguishable from Russians, but many Caucasians remembered the part that Cossacks had played in the original Russian conquest of the region.

Awkwardly for both Azerbaijan and Russia, the main railway line linking them, and the oil pipeline from Baku on the Caspian Sea to Novorossiysk on the Black Sea, both ran through Chechnya, and were cut during the war there. New discoveries of huge reserves of oil in the Caspian and around its shores were intensifying interest in the problem of getting it to outside markets. In 1993–6 a number of big deals were made by the region's governments with oil firms (both Russian and western); but in most cases their implementation was likely to need the consent of at least two governments. A new pipeline from Baku in Azerbaijan through Georgia and Turkey to the Mediterranean – avoiding Russia and bypassing both Chechnya and the congested Bosporus – opened in 2005. Future pipelines may bring central Asian oil and gas through the Caspian to the west, or to China or India. There was an ongoing dispute about ownership of the offshore oil. The Russians held that the Caspian was a lake, and its resources therefore the common property of all the surrounding states; others held that it was a sea, and that, under international law, each of those states could claim a national sector *(5)*. No comprehensive settlement of this question has been reached, but bilateral agreements among some of the five coastal countries (Russia, Azerbaijan, Iran, Turkmenistan, Kazakhstan) have allowed oil exploration to continue.

20 Ex-Soviet Central Asia

The breaking up of the Soviet Union produced five independent Asian countries, in the region once known as Turkistan. All five have Muslim (mainly Sunni) traditions and culture. Four use languages of the Turkic group – the fifth language, Tajik, being Iranian. The peoples of this central Asian region, conquered by the Russian empire between 1845 and 1895, broke away after the 1917 Russian revolutions, but by 1924 the USSR had reconquered them. It later reorganized the area into five Soviet republics. The division was mainly based on language, but exceptions to that rule were made in order to break up centres of potential resistance. One of these was the Fergana valley, which before 1876 had been the heart of the powerful khanate of Kokand, and even in the 1930s was notable for guerrilla resistance to rule from Moscow. Hence the curious intertwining in and around it of the boundaries, laid out in the Soviet period, which still exist today.

The USSR's large and fast-growing Asian population had caused increasing concern to its ruling circles. There were fears that these Asians would become restive on seeing other Asians winning independence in the era of European 'decolonization' *(27)*; that Islamic fundamentalism might get a hold, especially after the mullahs seized power in neighbouring Iran *(47)*; that there would be angry reactions to the sight of the Soviet army fighting Afghans, many of whom had kinfolk in central Asia *(50)*. The decision to withdraw the army from Afghanistan in 1989, and to abandon the Marxist rulers in Kabul to their fate, cannot have helped the USSR to retain the Asians' respect for its military power and political skill. Yet, after 1991, old-style communists were able to keep control in central Asia more successfully than in other ex-Soviet regions; and Russia found that it could, usually, still get its way in central Asia.

Kazakhstan had become the most Russified of the five republics. Millions of Russians and other Europeans (including ethnic Germans from the Volga region) were settled there, mainly in its northern areas. In 1989, Kazakhs were only 40%, and Russians 38%, of its population of 17 million. In 1996 it was decided to transfer the capital, by 1998, from Almaty (formerly Alma-Ata), in the south-east, to Astana (formerly Akmola and Tselinograd). There, in the north, it would be conveniently closer to Kazakhstan's main industries, its best farmland, its oilfields and goldmines; but the move was also seen as a sign of the government's anxiety to prevent a breakaway by the north, where the Russians were a local majority – and were unhappy about the loss of their pre-1991 privileges. Emigration continued to shift the ethnic balance in favour of Kazakhs, now 53% of the population, but Russian influence remained strong and Russia retained control over its space base at Baikonur.

Uzbekistan, the most populous state in the region (27 million people, only 6% of them Russians), contains Tashkent, central Asia's biggest city, and Bukhara and Samarkand (once the capital of Timur's empire), two ancient centres of Islamic civilization. Before the

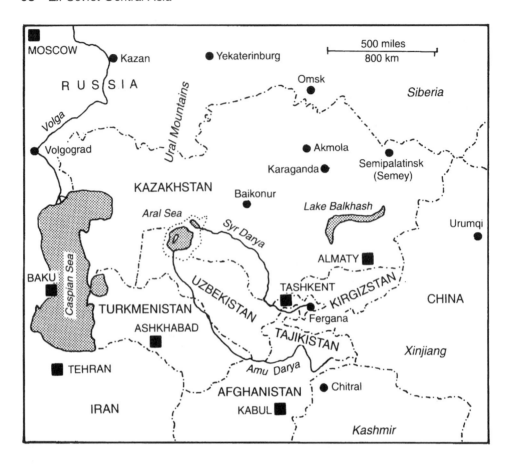

Russian conquest the Uzbeks had been dominant in central Asia; and after 1924, as a precaution against their resurgence, their republic was given boundaries that left large numbers of Uzbeks outside it. The Soviet period brought a great expansion of the cultivation of cotton, and this required irrigation on a scale that overstrained the arid region's water resources. Uzbekistan's other resources include gold and uranium. After Russia it is Asia's largest producer of uranium and it has long accounted for about 30% of gold production in the former Soviet region.

Turkmenistan has only 5 million inhabitants (4% Russian), and much of its area is desert; but here, too, irrigation made possible the large-scale production of cotton. Before 1991 it produced 11% of the USSR's natural gas; and it now pins great hopes on a massive increase in its exports of gas and of offshore oil from the Caspian – if the problems of getting these fuels to foreign markets can be solved *(19)*. Until his death in 2006, President Saparmurat Niyazov controlled the country for two decades and built a personality cult rivalling that of the Kims in North Korea *(59)*, naming himself and the month of January Turkmenbashi (Leader of the Turkmen) and requiring people to study his *Rukhnama* (Book of the Soul).

Tajikistan (with 7 million people, 1% Russian, 15% Uzbek) and *Kyrgyzstan* (5 million, 13% Russian, 14% Uzbek) are mountainous countries with very limited resources. In 1992, Tajikistan was torn by civil war; the opposing factions, often labelled communist and Islamist, also represented an antagonism between the country's western areas and its

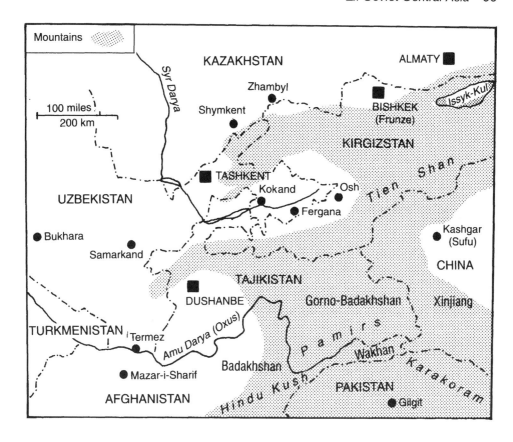

eastern highlands, which include the main part of the Pamirs. Although the highlanders were defeated, the government in Dushanbe failed to resolve the hostility. By 1993 it had called in 7,000 Russian soldiers, rising to 25,000 by 1996. Many were stationed on the Afghan frontier, across which bands of rebels still launched raids, with some help from fellow Tajiks in Afghanistan.

The Tajiks, the Uzbeks and the Turkmens in ex-Soviet central Asia all have kindred in Afghanistan; and after 1994, when the continued fighting in that country developed into an open contest for power between its main ethnic and linguistic communities *(50)*, there was an increased possibility of the central Asians becoming directly involved in it. However, there was a restraining factor in the presence of Russian forces along this part of the former Soviet frontier. In addition to the Russian soldiers in Tajikistan, there were 15,000 guarding Turkmenistan's borders with Afghanistan and Iran.

While closing bases elsewhere, in 2003–4 Russia opened its first new bases abroad since the Soviet collapse in Kyrgyzstan and Tajikistan, with the aim of fighting terrorism and monitoring Afghanistan and the drug trade.

All the republics except Kyrgyzstan were tightly controlled by former communists who might have changed their party labels but had not changed their hard-line methods. Opposition parties, if not always banned, were always harassed; newspapers and broadcasting were kept under state control. Presidents had their terms in office extended by referendums in which the published figures for 'yes' votes followed the old patterns: 99.99% in Turkmenistan in 1994, 99.96% in Uzbekistan in 1995.

After 1991, Turkey sought to gain influence in this mainly Turkic region; Iran and Pakistan also showed an active interest, and other Muslim countries offered money and support for the building of mosques and the restoration of Islamic monuments and institutions. The central Asian rulers welcomed such offers, and readily discussed the prospects of developing trade and economic co-operation with neighbouring states outside the former Soviet sphere; but on nearly all important issues they still looked mainly to Russia, with which or through which their countries still did most of their business. The habit of accepting guidance from Moscow seemed hard to break.

One direct legacy of Soviet 'planning' was posing a fearsome problem for them. The massive diversion, for irrigation of cotton-growing areas, of water from the Amu Darya (Oxus) and Syr Darya rivers had caused great environmental damage. Most alarmingly, it shrank and eventually split the Aral Sea into two parts as the water level continued to fall. The surface area of the sea fell by two-thirds from 1960 to the early 2000s, replaced by salty desert; large areas around it became infertile and unhealthy; the cotton crop itself was reduced. In 1993 the five republics promised to finance a programme to save the Aral Sea, but little money emerged.

China, too, sought better ties with Central Asia, to contain separatists in its restive Muslim regions and to gain access to new energy reserves. In 2001, building on earlier agreements, China, Russia and all of the states in the region except Turkmenistan formed the Shanghai Co-operation Organization (SCO). Officially, its purposes were to counteract 'terrorism, separatism and extremism' (by ethnic minorities and Islamists) and to improve economic ties. Russia and China both invested in the region's oil industry; oil pipelines from Kazakhstan to Russia (for export via the Black Sea) and China were completed in 2001 and 2005.

At least as important as strategic and economic ties was a desire to counter American influence in the region – initially by encouraging Uzbekistan, Tajikistan and Kyrgyzstan to close their US military bases. All three had supported the US-led war in Afghanistan in 2001, but in 2005 Uzbekistan told the Americans to leave what had been their most important base north of Afghanistan.

Iran, India, Pakistan, Mongolia and Afghanistan have attended SCO meetings; the inclusion of Iran in particular, and the SCO's potential as a military alliance, raised concerns outside the region.

21 Scandinavia

Traditional Scandinavian neutrality was shattered by the Soviet invasion of Finland in 1939 and Nazi Germany's occupation of Denmark and Norway in 1940. Britain prevented the Germans from extending their occupation of Denmark to Iceland, which became an independent republic in 1944, or to Greenland *(77)*. The Finns, hoping to regain the territory the USSR had seized, joined Germany in attacking it in 1941. Only Sweden remained neutral throughout the 1939–45 war.

After 1945 the Scandinavians sought to rebuild their grouping. A Nordic Council was set up, as a means of consultation between ministers and members of parliament from Denmark, Finland, Iceland, Norway and Sweden. A Scandinavian defence union was also discussed, but Denmark, Iceland and Norway, fearing that this would not provide enough security, chose instead to join the North Atlantic alliance in 1949.

Their fears arose largely from the Soviet Union's annexation of the three small Baltic states *(18)* and its grip on Poland and East Germany, which brought a large part of the Baltic Sea's shores under its control. It had also annexed Finland's only Arctic port, Petsamo (now Pechenga), and it built large naval, air and army bases in the Kola Peninsula. In 1948 it put pressure on Norway, and marred the Finns' neutrality by making them promise to help defend the USSR in certain circumstances. From time to time it renewed its pressure on Finland and, indirectly, on Sweden. In the 1980s, Sweden was troubled by Soviet submarines' furtive entries into its coastal waters (one ran aground near a Swedish naval base).

Soviet pressure on Finland blocked attempts to form a Nordic economic union. Denmark joined the European Community in 1973; Finland and Sweden joined in 1995. The Norwegians, in two referendums, voted to stay out. Iceland did not seek membership. (Both countries are within the European free trade area *[12]*.) When Denmark joined, its self-governing Faroe islands *(22)* stayed out; when the Danes gave Greenland self-government, Greenland withdrew from the EC.

The dramatic changes in the Soviet sphere from 1989 to 1991 transformed the situation in the Baltic region. The Nordic countries were deeply interested in, and greatly relieved by, the three Baltic states' recovery of independence. Collapse of the USSR was most difficult for the Finns, as the USSR had been their biggest trading partner. Among the many old links between Scandinavia and the Baltics, the closest were Finland's links with Estonia (the two languages have much in common).

Immigration has brought a substantial non-European population to the region for the first time in its history. In 2005 more than 10% of the Swedish population had been born elsewhere. Opposition to immigration, in particular that of Muslims from Asia and Africa, boosted the vote for right-wing parties in Denmark. A Danish newspaper in

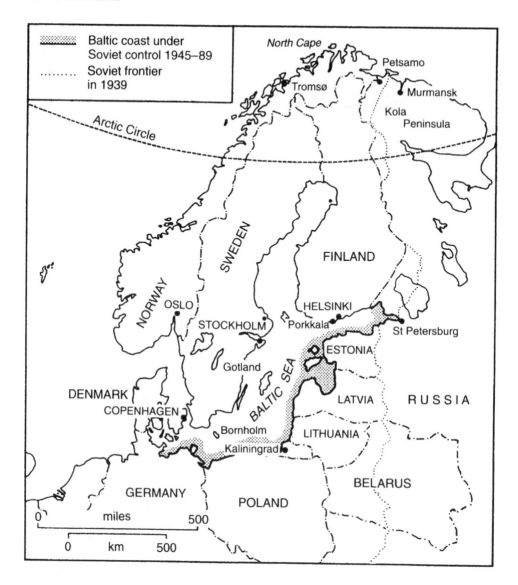

Baltic coast under
Soviet control 1945–89

Soviet frontier
in 1939

North Cape

Petsamo

Tromsø

Murmansk

Kola
Peninsula

Arctic Circle

SWEDEN

FINLAND

NORWAY

OSLO

HELSINKI

STOCKHOLM

Porkkala

St Petersburg

Gotland

ESTONIA

BALTIC SEA

DENMARK

LATVIA

RUSSIA

COPENHAGEN

Bornholm

LITHUANIA

Kaliningrad

BELARUS

GERMANY

POLAND

miles   500

km   500

2005 published cartoons depicting the prophet Muhammad; there followed riots and
dozens of deaths, attacks on European embassies, and boycotts of European products
around the Islamic world.

22 Northern Seas

The coastal states around the North Sea have shared out the rights to its seabed oil and gas by drawing dividing lines based mainly on equidistance from coasts, on the authority of the 1958 UNCLOS convention on the 'continental shelf' *(5)*. Britain – largely thanks to its Shetland and Orkney islands – and Norway got most of the oil. Later, Britain and France fixed a dividing line running down the Channel, with the British Channel Islands in a small sea enclave on the French side of the line. As to some of the Irish–British dividing lines, disputes were less easily resolved because Ireland rejected Britain's argument that baselines should be drawn from small islands, including the uninhabitable Rockall islet, which Britain had formally annexed.

The continental shelf is wide in this region; it underlies large sea areas around Britain and Ireland, as well as the whole of the North Sea except for a narrow band of deeper water along Norway's coast. But the 1958 shelf convention did not resolve disputes over fishing rights. In the 1950s and 1960s, Norway and Iceland, whose coastal fisheries had been much used by British and other trawlers, widened their fishing limits, setting off a series of disputes. In the 1970s, Iceland extended its limit to 200 miles, and had several (bloodless) 'cod wars' with Britain, whose fishing fleet was hit particularly hard by exclusion from Icelandic and Norwegian waters. A desire to retain control over its fishing stocks – the major source of export revenue – deters Iceland from seeking to join the EU.

Many other coastal states announced claims to mineral and fishing rights in 200-mile-wide 'exclusive economic zones' (EEZs) during the 1970s, without waiting for the new UNCLOS code of sea law, which did not come into effect until 1994 *(5)*. The European Community states proclaimed a joint EEZ. In the North Sea this zone roughly corresponded to all the 'oil sectors' combined, except Norway's. Britain, which 'contributed' the largest part of the joint zone and of the stocks of fish in it, objected to the idea of all Community members having unrestricted fishing rights throughout it. There was continuous wrangling among EC states over fishing quotas until 1983, when an agreement was reached. This did not settle all the disputes; and meanwhile the North Sea fish stocks were being ravaged by overfishing.

Farther north, when Norway proclaimed a 200-mile zone around its Jan Mayen island, disputes about dividing lines arose with Denmark, because of Greenland *(21)*, and with Iceland. Norway had a more serious dispute with the Soviet Union in the Barents Sea, which is rich in fish and may prove rich in oil. Here, while Norway wanted a median dividing line based on equidistance from coasts, the USSR argued that this was a case of an Arctic 'sector' boundary which should be based on lines of longitude.

The same two countries were also at odds over the seabed around Svalbard (Spitsbergen). In a 1920 treaty these islands were acknowledged to be Norwegian territory, but other

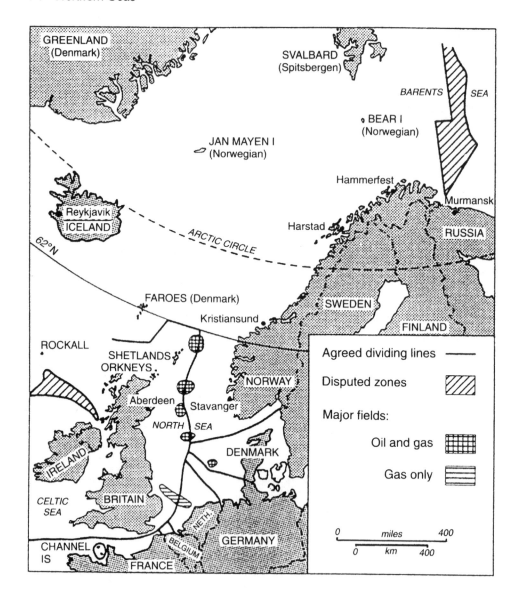

nations were given the right to exploit their resources on the basis of equal access under Norwegian law. Both Norway and the USSR have mined coal there. The Soviet contention was that all signatories to the 1920 treaty ought to have similar access to the resources of the islands' continental shelf. Norway argued that the Svalbard shelf should be treated as part of its own shelf.

In the late 1940s the Soviet Union had tried to make Norway let it establish a garrison in Svalbard. For some time after that, Norway kept a wary eye on the Soviet coal-mining camps for fear that a military presence might be created there. In the early 1990s some Soviet citizens were leaving Svalbard, but coal-mining continues, and Russians and Ukrainians are a third of the population. Gold was discovered in Svalbard in 2002.

The Soviet Union had strategic motives for discouraging other nations' activities in the Barents Sea. It wanted to keep foreigners well away from its naval bases around Murmansk in the Kola peninsula. The only way its warships and submarines could enter the Atlantic without having to pass through closely watched straits was by using these bases, which are ice-free – the effect of the Gulf Stream's continuation in the North Atlantic Drift being to extend ice-free water far north of the Arctic Circle in these northern seas *(77)*.

After the break-up of the Soviet Union, there was a marked reduction of tension in this region, and hopes that co-operation might replace confrontation. In 1993 the Barents Council was set up to promote peaceful development and the rights of indigenous communities such as the Sami ('Laplanders') in northern Finland, Sweden, Norway (not Svalbard) and north-western Russia, with Denmark, Iceland and the EU also joining as members.

23 Minorities and Micro-States

In western Europe, some separatist demands and disputes between language groups have led to violence, some only to political agitation – although, at times, extremist groups have resorted to violence on the fringes of non-violent movements. (See *14–20* for the minority problems of eastern Europe and the former Soviet Union.)

In *Belgium*, Flemish, a variant of Dutch, is spoken mainly north of a line running just south of the (bilingual) capital, Brussels, which hosts both many EU institutions and NATO headquarters. French is spoken mainly in the south, whose inhabitants are known as Walloons. French used to be dominant; but the Walloons are outnumbered, and economic power has shifted to the Flemish north. Since the 1970s a series of constitutional changes have made Belgium a loose federation, but some Flemings still voice separatist aims.

In 1993 a law gave Belgian courts 'universal jurisdiction' in cases involving genocide or war crimes – the right to pursue cases regardless of where the events occurred. This law was used to pursue people involved in the Rwandan genocide *(36)* and similar activities elsewhere, but was reduced in scope in 2003 following foreign protests at attempts to prosecute serving national leaders as well as the founding of the International Criminal Court (2002).

In *Britain*, the Channel Islands and the Isle of Man have long enjoyed autonomy, and Scottish and Welsh nationalists made headway in elections in the 1960s and 1970s. In 1979 referendums were held on proposals for devolution; the voting was negative in Wales and indecisive in Scotland. Those proposals lapsed, but nationalist agitation continued, the Welsh movement focusing on language rights, while some Scots had North Sea oil *(22)* particularly in mind. A second attempt at devolution in 1997 was more successful – referendum victories led to parliaments in 1999. Because the Scottish victory was much more decisive, the parliament for Scotland gained broader powers than that of Wales, including some control over taxation.

In *France* elected councils were created in all the 22 regions during the 1980s, after years of sometimes violent agitation in Corsica and Brittany; in the Languedoc area there had also been discontent with the traditionally centralized administrative system. In the 1980s some acts of violence were carried out by French Basques, who had been influenced by events across the border in Spain.

In *Italy* there is now a system of regional autonomy, but in the early 1990s the Northern League party demanded an almost complete separation for the industrialized north of the country, for which it coined the name 'Padania'. An older problem concerned the Alto Adige region – the provinces of Bolzano (Bozen) and Trento (Trent) – which before 1918 had been Austria's South Tirol. A majority of Bolzano's inhabitants speak German;

after an agitation that included some terrorist violence, their province was given autonomy in 1972.

Spain and *Portugal* retained fascist rulers into the 1970s. Their transition to democracy was accompanied by withdrawal from most remaining foreign colonies, including Mozambique, Angola, East Timor and Western Sahara. Migrants from its former African colonies have arrived in Portugal, while economic growth since Iberia's accession to

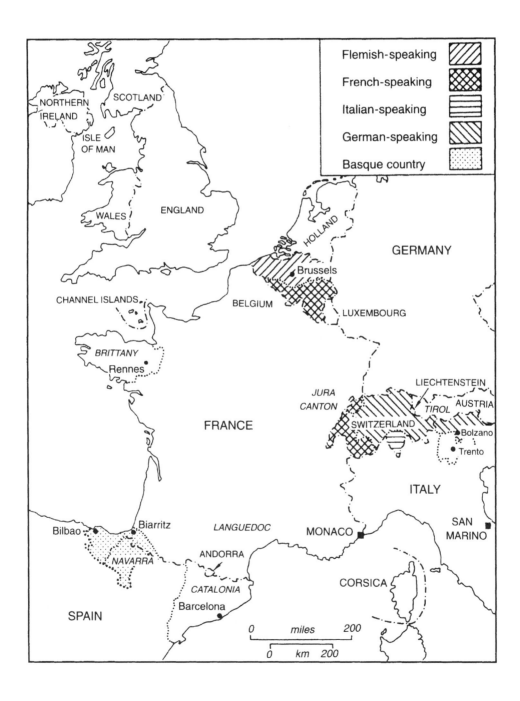

the EU in 1986 has attracted Moroccans and other Africans to Spain, sometimes via the two remaining Spanish enclaves in Morocco *(40)*. With the liberalization of intra-European migration since EU accession, coastal areas have attracted large numbers of retired northern Europeans – much as the southern US attracts American and Canadian retirees *(68)*.

In Spain old regional rights were suppressed during General Franco's dictatorship, and a violent separatist movement developed in the Basque provinces (Navarra is not classed as one of those provinces; its inhabitants are mostly of Basque origin but few speak Basque). After Franco's death in 1975 a plan for regional autonomy was adopted, starting with the Basque provinces and Catalonia; they elected assemblies in 1980. Tax revenue sharing influences both disputes – both Catalonia and the Basque area are among Spain's richest regions. Recently a third region, Valencia, has also claimed more autonomy. One Basque group, Euskadi Ta Askatasuna (ETA), demanded independence, but most Basques opposed its persistence in terrorism. Except for a ceasefire from 1998 to 2000, and the declaration of a 'permanent' ceasefire in 2006 (soon broken), ETA has carried out a sporadic bombing campaign – overshadowed in 2004 by the simultaneous bombing of several Madrid trains by Islamic terrorists.

Switzerland is a loose federation of 23 cantons. A majority of the Swiss are German-speakers; others speak French or Italian; 50,000 speak an old local language, Romansh. In 1978 a new French-speaking canton of Jura was created after long agitation in an area formerly included in the mainly German-speaking canton of Bern. Swiss women received the right to vote in federal elections only in 1971, but universal suffrage in local elections did not cover all cantons until 1990.

Switzerland has maintained a policy of neutrality in European conflicts. It has long hosted international organizations – including the largest UN office outside the New York headquarters – and has large migrant and expatriate populations. Nevertheless, it did not join the UN until 2002, has rejected EU membership, and retains strict citizenship criteria. In 1997, Switzerland's reputation as a financial centre was tarnished when a bank guard lost his job for preventing the destruction of documents relating to old dormant accounts owned by Jews murdered in the Holocaust. In fleeing Switzerland to avoid arrest, the guard became the first Swiss to be granted political asylum by the US, and several years later Switzerland agreed to set up compensation funds for Holocaust victims.

Luxembourg, a grand duchy with 500,000 people, uses three languages: French, German and its own Letzeburgesch. The map shows several even smaller sovereign states, with populations of between 30,000 and 80,000: *Andorra, Liechtenstein, Monaco* and *San Marino*; there is also the tiny *Vatican City* state in Rome. Each micro-state has economic and other links with larger neighbours: Luxembourg with Belgium, Andorra with France and Spain, Liechtenstein with Switzerland, Monaco with France, San Marino and the Vatican state (the Holy See) with Italy.

Since 1945, large-scale immigration into western Europe from other regions has created minority problems of a new kind. The new immigrant communities, although too dispersed for marking on a small map, tend to be concentrated in big cities. Notable among them are: in Britain, those from southern Asia (India, Pakistan and Bangladesh), the West Indies and eastern Europe; in Holland, refugees from Surinam, Ambonese from Indonesia *(64, 73)*, Turks and Moroccans; in Germany, Turks, Kurds, and refugees from the wars in ex-Yugoslavia; in France, and to a lesser extent in Italy and Spain, Arabs from North Africa *(39)*. One result of the new migration has been the growth of large Muslim communities in western Europe, particularly in France, Germany, Holland and Britain. Muslims are younger and more religious than other Europeans, whose support for immigration restriction has been enhanced by terrorism and violent reaction to critics of Islam in Spain, Britain, Holland and Denmark.

24 Ireland

The state that is called Britain, for short, in this book is formally the United Kingdom of Great Britain and Northern Ireland. It was the United Kingdom of Great Britain and Ireland from 1801 until, after a bloody guerrilla war, Ireland was partitioned in 1921. The treaty gave effective independence to the greater part of Ireland – which became a republic in 1949 – while six counties in the province of Ulster remained in the UK, with a 'home rule' system and a Northern Ireland parliament at Stormont in Belfast. But the border was kept open; and Irish migration to Britain continued.

The republic's constitution asserted a claim to the whole island. In the north, the Protestant majority's votes in every election showed their opposition to the idea of joining the republic, which is now almost entirely Catholic (after 1921, many Protestants left) and whose main traditions are of resistance to British rule. At times, the Irish Republican Army (IRA) and other groups carried out terrorist actions in the north, using bases in the south – whose government, at times, dealt with them severely.

In the 1960s the prime ministers in Dublin and Belfast joined in trying to improve relations, with support from Britain; but a backlash in 1969 led to violent clashes in northern cities. The local police failed to protect Catholics against attacks by Protestant mobs, and British troops were sent to restore peace. A revival of the dormant IRA followed. Strikes by Protestant workers blocked the creation of a 'power-sharing' system and of a north–south 'Council of Ireland'. 'Home rule', which had come to mean permanent Protestant rule from Stormont, was ended in 1974.

Direct rule from London improved the Catholics' position; and in 1985 Britain promised to give the Dublin government some say in the north's affairs. The Protestants' fears increased (emigration was reducing their majority); IRA terrorism was countered by that of 'loyalist' groups. Although the rate of sectarian killings in Northern Ireland fell off (the annual average was 260 in 1971–6, 80 in 1977–94), more Catholics than Protestants were being killed there in the early 1990s. However, IRA gangs were also murdering people in Britain and elsewhere. (They made two attempts, in 1984 and 1991, to kill British prime ministers.)

In 1994 both the IRA and 'loyalists' announced ceasefires. But the IRA resumed its planting of bombs – in Britain at first – in 1996. Although the 'loyalists' still maintained their ceasefire, they warned that they would not do so much longer unless the IRA and other 'republican' extremists showed similar restraint.

A more lasting peace agreement was finally approved by voters in both Northern Ireland and the Irish Republic in 1998. In 2005 the IRA agreed to disarm, but its conversion to an entirely political organization remains incomplete and the new Northern Irish parliament established by the 1998 agreement has not met since 2002.

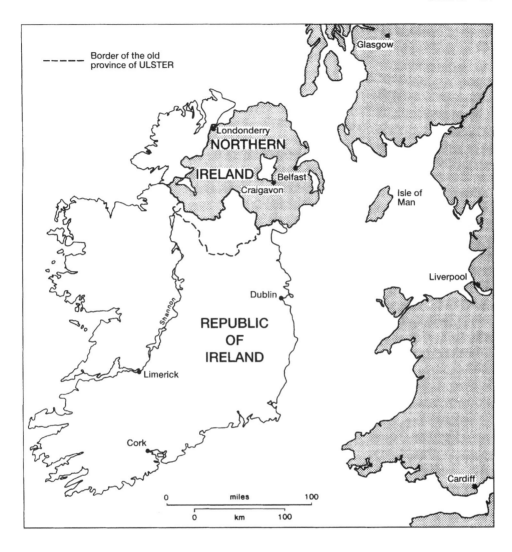

A higher Catholic birth rate, emigration and a general decline in identification with organized religion has eroded the Protestant numerical advantage. In the 2001 census, 40% of Northern Irish identified themselves as Catholic, while 45% chose a Protestant denomination.

More than any other country, the Irish Republic has benefited economically from joining the EU. Long one of Europe's poorer countries, Ireland exported people to Britain, America and Australia for more than a century following the famines of the 1840s – the current population of 4 million is 40% below the mid-nineteenth-century peak. With export-oriented foreign investment and EU aid, it now leads the EU in per capita GDP (except for Luxembourg), and in the mid–1990s the migration balance became positive for the first time in 150 years.

25 Gibraltar

In 1996, Spain put pressure on Gibraltar by obstructing frontier crossings. It was copying the tactics previously used, without success, by General Francisco Franco, who, after starting an army mutiny and winning the subsequent 1936–9 civil war with the help of German and Italian forces, had ruled Spain for 36 years.

Gibraltar was captured by Britain in 1704 (and later renounced by Spain by the Treaty of Versailles, in exchange for Florida and Minorca). It has been held by Britain longer than by Spain, for Spain had not taken it from the Moors until 1462. For two centuries it was a key British naval base; it resisted a siege by French and Spanish forces from 1779 to 1783, and it showed its usefulness at the time of the great naval battles at Cape St Vincent (1797) and Trafalgar (1805). When the Suez route was Britain's imperial lifeline, Gibraltar and Malta were important stages along it *(42)*. Malta became independent in 1964; but independence did not seem a practical aim for Gibraltar. Although its population is as large as that of such micro-states as San Marino *(23)*, it is only a four-mile-long peninsula, overshadowed both by the 1,400-foot Rock and by Spain's recurring claims.

In 1964, when the Gibraltarians were given fuller internal self-government, General Franco began to try to force them to accept his rule. His regime banned trade with Gibraltar and obstructed frontier crossings, stopping them completely from 1969 onwards. The Gibraltarians were not cowed. In a 1967 referendum they voted almost unanimously to retain their links with Britain; and they confirmed this choice by their voting in each subsequent election for their House of Assembly. In 2002 another referendum found 99% opposed to Britain and Spain sharing control of the territory.

Meanwhile, the new 'siege' imposed by Spain hit the economy of the adjacent Spanish region (the 'Campo') harder than that of Gibraltar. It continued after Franco's death in 1975, mainly because Spanish army officers were able to put pressure on Spain's elected governments. The frontier was not fully reopened until 1985.

Spain argued that it could not tolerate a foreign enclave on its coast. However, Spain continued to hold two 'autonomous cities' on the coast of Morocco: Ceuta and Melilla, the last places in Africa still ruled by a European state *(40)*. Morocco indicated that it would raise the question of these two enclaves if Gibraltar were transferred to Spanish control.

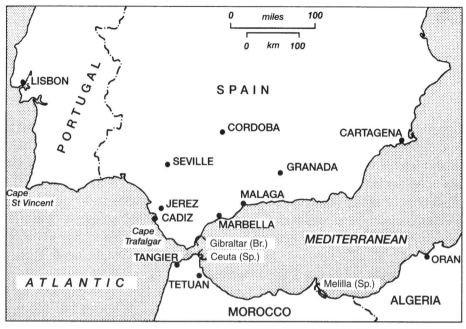

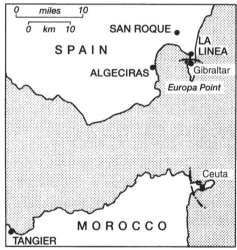

26 Cyprus, Greece and Turkey

Cyprus was ruled by Turkey from 1570 to 1878, then by Britain, but three-quarters of its population (now 800,000) is Greek. When it became independent in 1960, Britain, Greece and Turkey were given the right to keep forces in Cyprus, and to intervene to maintain its constitution, which included provisions designed to protect the Turkish minority. At independence, Britain retained sovereignty over bases at Akrotiri and Dhekelia.

By 1964, Greek pogroms had driven the Turks into small enclaves (including part of the capital, Nicosia). A United Nations force was sent to forestall a Turkish invasion. The UN men kept conflicts from escalating until 1974, when the army officers then ruling Greece organized a coup in Nicosia, aiming at '*enosis*' (union of Cyprus with Greece). Turkey sent troops to occupy the north of the island; 180,000 Greeks fled to the south, 45,000 Turks to the north. In Greece the 'rule of the colonels' collapsed, and democracy was restored.

Many unsuccessful attempts at UN mediation in Cyprus were made. Turkey kept 30,000 soldiers there, and the 'Turkish Republic of Northern Cyprus' – recognized only by Turkey – was proclaimed in 1983. By 1996 the UN force, reduced from 6,000 men to 1,200 (mainly Argentine and British), could not always prevent clashes between Turkish soldiers and gangs of Greeks attempting to cross the buffer zone. The Greek-Cypriot National Guard (which Greece supplied with officers) made no effort to stop such attempts.

Conflicting claims in the Aegean Sea brought Greece and Turkey close to war in 1976, 1987 and 1996. Greece rejected Turkey's claim to part of the 'continental shelf'; Turkey wanted access to some of the Aegean's oil potential, and showed that it would reject a claim to 12-mile territorial waters for the Greek islands close to Turkey's coast *(5)*.

These quarrels led Greece to veto Turkey's requests for economic links with the European Union, and to obtain an EU ban on trade with northern Cyprus. In 1995 the EU agreed to start talks about the admission of (Greek) Cyprus to the union; Greece then lifted its veto, and a customs union between Turkey and the EU states took effect in 1996.

In 2002 the EU invited Cyprus to join the EU two years later. UN-sponsored talks sought to reunify the island prior to admission, and a 2003 referendum proposed a loose federation of north and south, with some territorial transfer to the Greek side and Turkey retaining the right to station troops in Cyprus. Turkey, itself pursuing EU membership *(12)*, supported unification, as did a majority of Northern Cyprus voters. But, with EU membership assured, the Greek side soundly rejected the plan, considering that it conceded too much to Turkey.

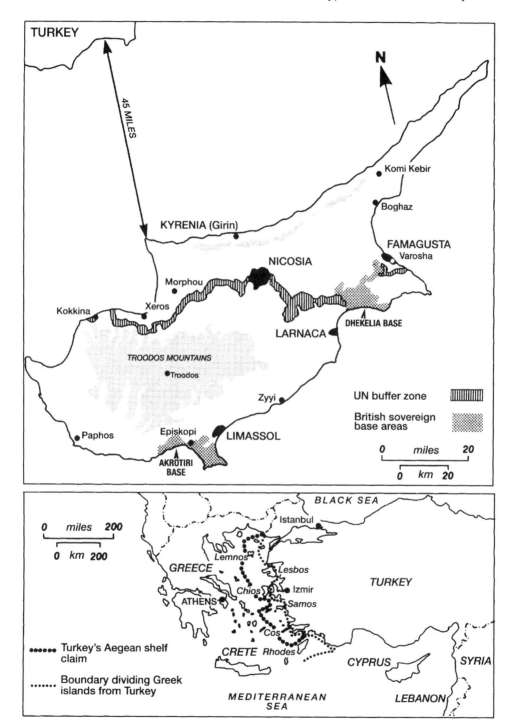

27 Asia and Africa

The map of these two continents has been transformed since 1945 by a rapid process of 'decolonization'. Before the 1939–45 war there were only a dozen sovereign states in the Afro-Asian world; now there are more than a hundred. Few vestiges remain of the colonial empires that European countries created after the first Portuguese penetration of the Indian Ocean region in the 1490s. In some places their ending followed long and bloody conflicts; but this uniquely big and swift change in the world map was also distinguished by the fact that, in most cases, the transfer of power was made by negotiation and without war.

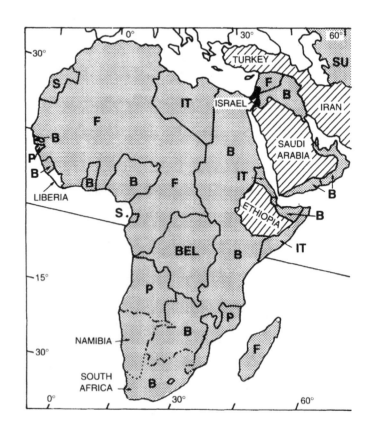

The greater part of the transformation took place within a period of only twenty years, starting in 1946–9, when the Philippines, India, Pakistan, Ceylon, Burma, Indonesia and several Middle Eastern states became independent. Portugal, the pioneer of colonization in Asia and Africa, was particularly reluctant to give up; it did not withdraw from any of its territories until after its own revolution in 1974. But since 1977, when France made Djibouti independent, the only places in Africa ruled from Europe are Ceuta and Melilla, Spain's two enclaves on the coast of Morocco *(40)*.

In Asia, only one large area was still under European control in the 1980s: the eastern part of the Soviet Union. The inhabitants of its Siberian northern regions are mainly Russian; and, although those regions are geographically Asian (they are close neighbours, on their Pacific coasts, to China, Japan and Korea), it was evident that they would remain part of Russia when the USSR broke up in 1991. To the south, in what had been Soviet Central Asia (and, earlier, Turkistan), five independent Asian states emerged; and a sixth state, Mongolia, gained an independence which it formerly had only in name *(20, 54)*.

Afro-Asian summit conferences were held at Bandung, in Java, in 1955 and at Algiers in 1965, but the second was a fiasco and no permanent grouping was formed. Instead, most of the Asian and African states joined the 'non-aligned' movement *(6)*. For many years, they could generally unite in pursuing the natural aims of such new states: primarily, to see that decolonization was completed and white-minority rule in South Africa ended. Once those aims were achieved, Afro-Asian unity became more elusive – although there remained a common interest in pressing richer 'northern' countries for more economic aid and for concessions in regard to trade.

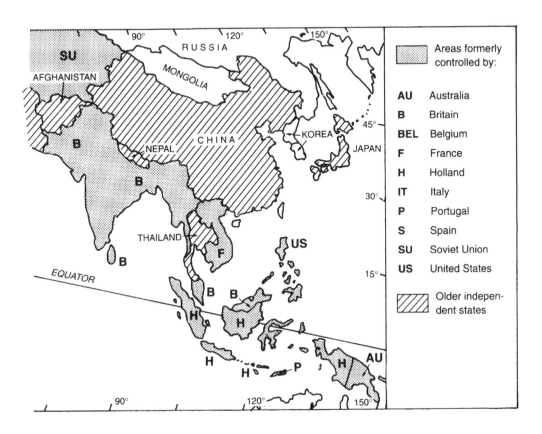

When 'east–west' rivalries impinged on their regions, Asian and African states some-times suffered, but sometimes profited, often contriving to play one side against the other. Complications faced them when China and the Soviet Union took to siding with contestants in the Afro-Asian world – China supporting Pakistan and Cambodia in their conflicts with Soviet-backed India and Vietnam. Later, when the Soviet Union and the 'three worlds' pattern disappeared, Asian and African states, like others, had to work out new approaches to a changed situation *(6)*.

China, as the biggest and strongest of the Afro-Asian countries (but never one of the 'non-aligned'), has posed some special problems for them. At times, it proclaimed itself their natural leader and champion; yet it invaded India in 1962 and Vietnam in 1979, and in the early 1990s its claims to islands in the South China Sea alarmed the South-East Asian states *(53, 55, 61)*.

Chinese economic expansion *(57)* has contributed to its military strength, but China has not (yet) sought worldwide influence on the scale of the Soviet Union. Thus, a new division has emerged between countries that align themselves (formally or tacitly) with the US and its allies and those that resist US influence – without being able to count on the support of a major power.

As for Japan – whatever the map may suggest – it cannot be counted as just one of the Afro-Asian countries. Some of those countries still show a certain reserve towards Japan because they have not forgotten its misdeeds in its years of aggressive expansion before 1945. More significant today, however, is the plain fact of Japan's status as an economic superpower with 'western' living standards *(2, 58)*. Inevitably, its interests and viewpoints have very little in common with those of a 'developing' (that is, still poor) country. As Asia's 'newly industrialized countries' (NICs) *(55)* climb towards the Japanese level, it can be seen that similar gaps – in outlook as well as in income – are widening between them and the poorer Afro-Asian states *(2)*. Yet it cannot be assumed that these newly enriched countries will adopt 'western' attitudes in all respects. As is already being seen, Asian societies can adapt remarkably well to changing circumstances without losing their familiar characteristics.

28 Islam

The Islamic religion took shape in Arabia nearly 1,400 years ago, and it was first spread by Arab conquests. Its scriptures are in Arabic. Mecca, in Saudi Arabia, is its holiest shrine, to which vast numbers of Muslims make a pilgrimage (*haj*). But only a fifth of the world's 1.3 billion Muslims are Arabs. There are more Muslims in India, Pakistan and Bangladesh than in the Arab countries. And not all Arabs are Muslim.

In historical perspective, the power of Islam was at its greatest height 300 years ago, when Muslim rulers controlled the Balkans, Greece, part of Russia and nearly all of India. It was in eclipse by the 1920s, after the breaking up of the old Turkish empire; there were then only five independent Muslim states. But since 1945 there has been a new increase in Islam's significance in world affairs. Many of the states that emerged from the old European empires in Asia and Africa *(27)* are peopled and ruled by Muslims; so are some of the states that have emerged from the former Soviet Union *(20)*; and several Muslim states are now major exporters of oil *(41)*.

In some Muslim countries (first and foremost, in Turkey; later, notably, in Iran) the twentieth century saw a movement towards religious toleration and a separation of religious authority from government. Now, in many places, a backlash against this secularism has brought a surge of Muslim fundamentalism; modernizing policies are branded as alien 'westernizing' ones. A dramatic example was the 1979 revolution in Iran *(47)*.

The Islamic Conference Organization (ICO), founded in 1969, had 57 members by 2001. They had been united, at least in principle, on such things as support for the Palestinian Arabs, for Afghanistan during the Soviet invasion, and for the Muslims of Bosnia *(15)*. But some members were alarmed when others – e.g. Iran, Iraq and Libya *(39)* – used Islamic appeals to disguise their ambitions.

There are sectarian divisions in Islam, as in other religions. From its earliest years the Shias (about 200 million) have been in conflict with the orthodox Sunni majority. For 500 years Iran has been the main Shia stronghold, and the mullahs who have ruled it since 1979, although posing as champions of all Islam, have exacerbated its divisions. Iran has worked hard to manipulate Lebanon's Shias and more recently those of Iraq, where Shias are in the majority. During its war with Iraq it tried, but failed, to win over the Iraqi Shias; renewed efforts followed the fall of the Sunni-led government of Saddam Hussein in 2003. Most Shias live at the heart of the Muslim world, with few east of Pakistan or west of the Red Sea.

In Britain in 1989 an author had to go into hiding when an imam in Iran, accusing him of blasphemy, told Muslims to murder him. This *fatwa* (Islamic legal ruling) against Salman Rushdie is still in force. Some Muslims living in Britain openly supported it; in other countries, two people involved in publishing Rushdie's work were killed. Iran's agents have also murdered a number of Iranian exiles in several European countries.

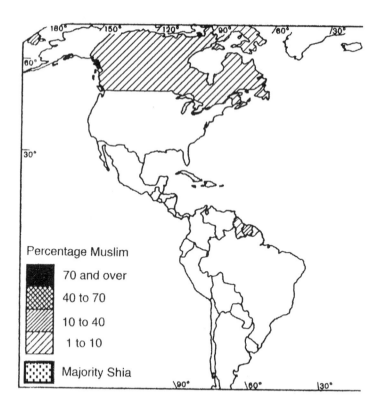

Neither the Iranians nor the Shias have a monopoly of modern Islamic fanaticism and fundamentalism. Libya's rulers have used fanatical assassins in Britain and many other countries. Sudan harbours Islamic extremist groups from other African countries, providing them with arms, money, and training in terrorist tactics. More than 50,000 lives have been lost in five years of conflict between Algeria's government and its fundamentalists, whose terrorism has forced most foreigners to flee the country *(39)*. Attacks on foreign visitors by Islamists in 1992–4 halved Egypt's earnings from tourism (which, however, revived in 1995 after a crackdown on extremists). A 1993 meeting of the Supreme Council for Islamic Affairs, attended by 35 Muslim countries' ministers of religious affairs, felt obliged to denounce terrorism and extremism as 'distortions that are alien to Islam'.

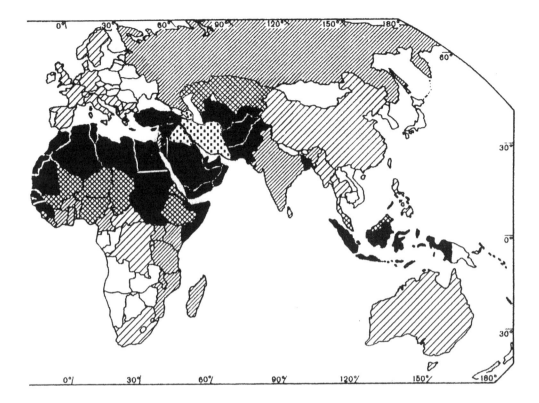

The idea of a holy war (*jihad*) against infidels is as old as Islam; but some modern rulers have twisted its meaning to suit their ambitions. Iran proclaimed that its war with Muslim Iraq from 1980 to 1988 was a *jihad*, and that its 700,000 dead were 'martyrs', killed while trying to 'liberate' Shia shrines at Karbala and other places in Iraq. In turn, Iraq claimed to be engaged in a *jihad* in the period 1990–1, when it seized Muslim Kuwait and attacked Saudi Arabia, the guardian of Mecca; but it was opposed and eventually defeated by a coalition that included forces from eight Muslim countries. In 1998, Osama bin Laden's 'World Islamic Front for Jihad against Jews and Crusaders' called on Muslims to kill Americans. Bin Laden, a Saudi, opposes the US military presence in Saudi Arabia, and his al-Qaeda organization is held responsible for several terrorist attacks *(8)*.

29 The Arab World

The Arabic language links 250 million people inhabiting a belt that runs from the Atlantic to the Indian Ocean. Most Arabs are Muslims (though there are large Christian communities in Egypt and Lebanon), and Mecca, Islam's holiest place, is in Arabia; politicians and rulers seeking pan-Arab support often exploit Islamic feeling. But when a sense of Arab unity has been strong it has been a reaction against alien rule – first by Turkey and then by west Europeans *(27)* – and, more recently, against the creation of the Jewish state of Israel near the centre of the Arab world. Reaction to the latter included persecution of the region's large Jewish communities, and Arabic-speaking refugees from North Africa, Iraq, Syria and Yemen formed a large part of Israel's population by 1960.

The Arab League, founded by Egypt, Iraq, Jordan, Lebanon, Saudi Arabia, Syria and Yemen in 1945, was later joined by Algeria, Bahrain, Comoros, Djibouti, Kuwait, Libya, Mauritania, Morocco, Oman, Qatar, Somalia, Sudan, Tunisia, the United Arab Emirates and the Palestine Liberation Organization. Its headquarters was moved from Cairo to Tunis in 1979, when Egypt's membership was suspended, but returned to Cairo when Egypt was readmitted in 1989 *(44)*.

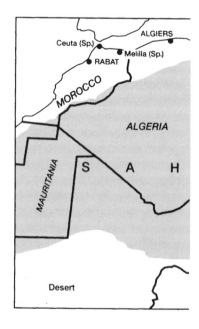

No clear line marks the Arab world's borders. The south Sudan is neither Arab nor Muslim. Djibouti and Mauritania are only partly Arab, and Somalia is not usually regarded as an Arab state. Iraq's north is peopled by Kurds *(49)*; Iran's south-west partly by Arabs *(47)*. The Berber language, which was widely spoken in North Africa before the eighth-century Arab conquest, survives in parts of Algeria and Morocco.

Seven small Gulf states formed the United Arab Emirates (UAE) in 1971–2. Yemen and the former South Yemen (Aden) were united in 1990. Other attempted mergers have failed. The United Arab Republic (UAR) formed by Egypt and Syria in 1958 was dissolved in 1961.

Sub-regional groupings have appeared. In 1981, Saudi Arabia and five small neighbours created the Gulf Co-operation Council *(47)*. In 1989 the Arab Maghreb Union was formed in North Africa. But the Arab Co-operation Council formed by Egypt, Iraq, Jordan and Yemen in 1989 broke up in 1990 when Iraq invaded Kuwait; Egypt, with other Arab states, denounced Iraq and contributed troops to the line-up against it.

A shared language, and some amount of shared national feeling, has not prevented frequent quarrels. Disputes between Morocco and Algeria, Libya and Egypt, Saudi Arabia and Yemen, and Syria and Iraq have been particularly bitter. But Iraq's swoop on Kuwait was the first attempt by one Arab League member to seize and hold another one by force.

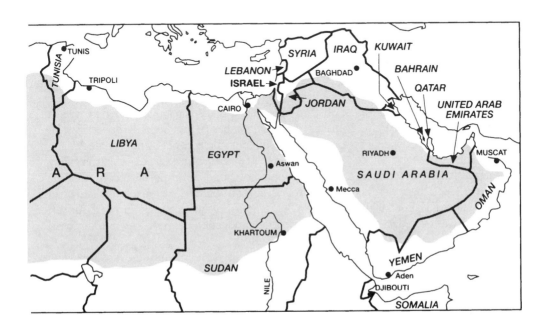

30 Africa

In Africa, European colonization and decolonization were both rapid *(27)*. Until the 'scramble for Africa' in the 1880s, Europeans had controlled only a few coastal strips and some areas in the extreme south and north *(34, 39)*. Then, within one century, nearly all of Africa came under European rule and re-emerged as independent states. Long guerrilla wars preceded the liberation of Algeria, Angola and Zimbabwe (Southern Rhodesia); but most of the other states attained independence peacefully. In 1977 it came to Djibouti, the last area ruled from Europe in 'black' or sub-Saharan Africa (that is, excluding Arab North Africa: *[39, 40]*). Rule by white minorities was ended in Zimbabwe in 1980, in Namibia in 1990, in South Africa in 1994.

Africa's deserts, dense tropical forests and lack of navigable rivers impede communications. Several hundred languages are spoken, and old tribal antagonisms have troubled many states. All except Morocco (which withdrew in 1984 when Western Sahara, under its control since 1975, was admitted as a member) are among the fifty-three members of the African Union, founded as the Organization of African Unity in 1963. As such, they usually discourage secessions and frontier changes. Although today's frontiers are mostly those that colonial powers drew – splitting or combining previously separate ethnic or linguistic communities – it is feared that failure to respect them would lead to much more fragmentation.

Recent conflicts have driven millions of Africans across frontiers and into refugee camps. In the 1980s and 1990s, droughts and civil wars combined to produce severe famine in several countries, notably Ethiopia, Mozambique, Somalia and Sudan. Government policies forcing migration or land redistribution magnified the suffering, as in Ethiopia in the 1990s and Zimbabwe since 2000.

As a whole, sub-Saharan Africa, which in the 1960s was keeping food production abreast of population growth and slowly raising its living standards, had become poorer by the 1980s, and acutely dependent on food aid from America and Europe. By then some African governments had begun to see the damage done by their own policies, which in many places had sharply discouraged the production of both food and exportable commodities. In the early 1990s there were signs of recovery in several countries where reforms had been initiated, but even after 2000 per capita GDP was still growing slowly in most African countries (it was even declining in a few), and food production per head had not improved since 1960.

To maintain a constant standard of living, a country's economy must grow as fast as its population; countries with high population growth struggle to maintain the even higher economic growth rate required to raise per capita GDP. In several African countries, the rate of population increase exceeds 3% per year – more than twice the world rate.

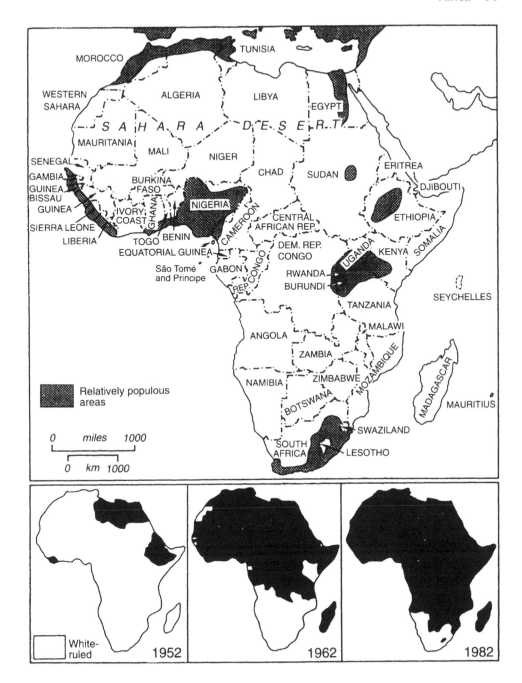

The burden of disease is also greater than elsewhere. Worldwide, 1% of people aged 15 to 49 are infected with HIV, the virus that causes AIDS; south of the Sahara the proportion rises to 7%, and it is above 20% in southern Africa. While not as widespread as malaria, AIDS kills adults in their working years; thus, its economic consequences are especially damaging.

Among African countries, the fastest growth rates in recent years have been found in countries recovering from wars (such as Angola and Mozambique) and where oil has been discovered (Equatorial Guinea, Chad). Democracies and other states with stable governments – Botswana (the world's largest diamond-producer), Cape Verde, Mauritius, Ghana, Uganda and a few others – have also managed consistent growth. War, famine, disease and dictators have disproportionately affected the rest.

31 Southern Africa

By the early 1960s European rule had ended in most of Africa north of the equator; but it continued in the south, a region strongly influenced by the white-ruled Republic of South Africa (RSA). The mineral wealth of southern Africa (including coal, cobalt, copper, diamonds, gold and uranium) was one cause of the creation of white communities. Another was that, even near the equator, there were healthy highland areas where commercial crops (e.g. coffee and tobacco) could be grown.

In 1963, Kenya became independent; a white-backed secession in Zaire's mineral-rich Katanga province (now Shaba) was ended by United Nations action *(32)*; and Britain dissolved the Central African Federation that it had created in 1953. The federation had united Nyasaland, Northern Rhodesia and Southern Rhodesia – today's Malawi, Zambia and Zimbabwe. Malawi and Zambia became independent in 1964.

Southern Rhodesia, nominally a British colony, was in practice ruled by its white minority. The original white settlement there in the 1890s had been made from South Africa and masterminded by Cecil Rhodes, who had made a vast fortune from South Africa's diamonds and gold. In 1965 the white minority proclaimed an independent Rhodesian republic. Through the United Nations, Britain had trade embargoes and other 'sanctions' imposed; but support from the RSA helped the white rebel regime to hold out against international pressure and African guerrilla resistance until 1980, when an election, held under British and Commonwealth supervision, produced a black government. The country's independence was then recognized, and it was renamed Zimbabwe (from a local site of ancient ruins *[32]*).

Portugal's revolution in 1974 (set off by soldiers weary of fighting long guerrilla wars in Africa *[33]*) brought independence for Angola and Mozambique in 1975. In their subsequent civil wars, one faction in each country was for years supported by the RSA. During the 1980s the RSA came under growing pressure to grant independence to Namibia, and by 1990 this had been done *(33)*. In 1994 white-minority rule was ended in the Republic of South Africa itself *(34)*, and it joined the Southern African Development Community. In the 14-member SADC, originally formed by states worried about the old RSA's strength, the new member was looked to, hopefully, as the whole region's economic powerhouse.

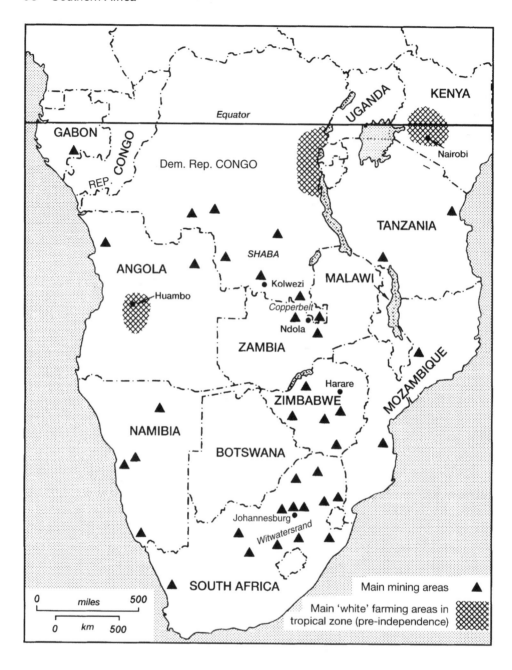

Main mining areas ▲

Main 'white' farming areas in
tropical zone (pre-independence) ▨

32 Central Africa

Congo-Kinshasa has only a short sea coast; Zambia and Zimbabwe are landlocked. Exploitation of their minerals *(31)* led to the building of rail links through Angola, Mozambique and South Africa; while those areas remained under white rule, the inland states' dependence on these links was of special importance. In the 1970s the Tazara (Tanzania–Zambia Railway) line was built, with Chinese help, to give Zambia an alternative outlet to the sea.

The former Belgian Congo was granted independence in 1960 hastily and without preparation. Its soldiers mutinied, and anarchy spread. Belgium sent troops to protect the 100,000 whites; a United Nations force was then sent to help to restore order and secure the Belgians' withdrawal. A secession of the mineral-rich Katanga province (now Shaba) was backed by various Belgian interests, with Rhodesian and South African support; the central government disintegrated, and a rival Soviet-backed regime was installed at Stanleyville (Kisangani). The UN helped to bring together a new national government and, in 1963, to end the Katanga secession.

Congo – renamed Zaire in 1971 – was plagued by more rebellions and mutinies (and invasions by ex-soldiers who had fled into Angola). In 1964, 1978 and 1991 it had to call in Belgian and French troops. By the early 1990s its economy was a wreck, its government a scandal under the dictator Mobuto Sese Seko, its army good only for extortion and looting.

In 1996, in its Kivu region, there was a new rebellion, in which the Banyamulenge, a people of the same origin as the dominant Tutsi minorities in neighbouring Rwanda and Burundi, were prominent. Helped by Rwanda's Tutsi army, the rebels captured Bukavu, capital of South Kivu province. Thus, at first, the fighting in eastern Zaire involved an extension of the struggle in Rwanda and Burundi between their Tutsis and their Hutu majorities *(36)*. However, it became clear that Zaire's troops would not fight and that the paralysed government in Kinshasa could not recover control of the eastern regions. The rebels gained widespread support as they advanced. Their capture of Kinshasa in 1997 forced the dictator into exile, and the country became Congo once again.

In 1998 rebels backed by Rwanda and Uganda once again took control of the east *(36)*, but with intervention from Angola, Namibia and Zimbabwe the new government survived. A peace agreement in 2003 ended a war that had cost at least 3 million lives, although smaller-scale conflict continued in the east. An election was held in 2006.

Even without war and the preceding decades of dictatorship and colonial rule, Congo's continued existence as a single country would be impressive: a population of 60 million people speaking perhaps 200 languages lives in an area more than half the size of the EU (and half rainforest), with little reliable transportation.

Zambia and *Malawi*, for some years after becoming independent in 1964 *(31)*, faced difficulties arising from their dependence on outlets to the sea through a still

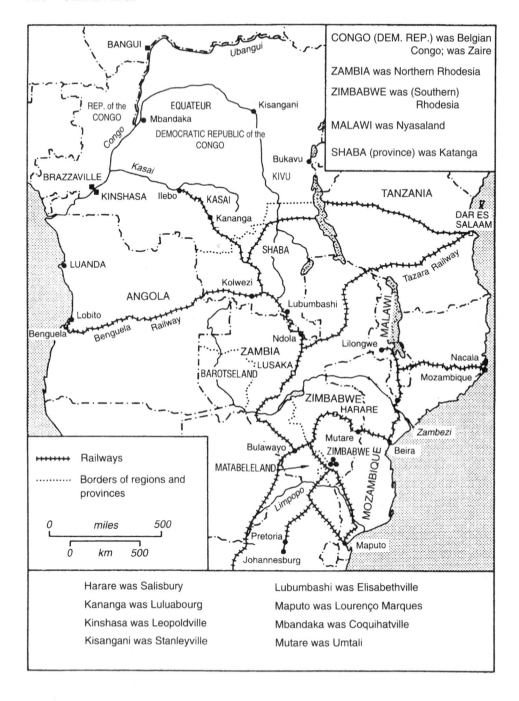

CONGO (DEM. REP.) was Belgian Congo; was Zaire

ZAMBIA was Northern Rhodesia

ZIMBABWE was (Southern) Rhodesia

MALAWI was Nyasaland

SHABA (province) was Katanga

Railways
Borders of regions and provinces

0   miles   500

0   km   500

Harare was Salisbury

Kananga was Luluabourg

Kinshasa was Leopoldville

Kisangani was Stanleyville

Lubumbashi was Elisabethville

Maputo was Lourenço Marques

Mbandaka was Coquihatville

Mutare was Umtali

Portuguese-held Mozambique – and, in Zambia's case, through a still white-ruled Southern Rhodesia (now Zimbabwe). African guerrillas fighting in Southern Rhodesia used bases in Zambia, which were then attacked by Rhodesian forces; the African National Congress (ANC), banned in South Africa, made Lusaka its headquarters; the imposing of sanctions (mainly trade embargoes) on Southern Rhodesia and South Africa created acute problems for Zambia. After 1975, when Portugal's withdrawal from Mozambique was followed by civil war there, nearly a million Mozambicans fled into Malawi.

Most of Zambia's export revenue still comes from copper. Mine output peaked in the 1970s when prices were high (in real terms), but fell in the 1980s when the price declined below the cost of production. By the 1990s prices rose again, but the lack of investment in government-owned mine infrastructure meant that production continued to fall. Privatization led to increased production after 2000, and Asian demand boosted prices through 2006.

Zimbabwe has been ruled by Robert Mugabe and his ZANU-PF (Zimbabwe African National Union-Patriotic Front) party since independence in 1980 *(31)*. Through the mid-1990s, despite antagonism between its two main tribal groups, the Shona majority and the once dominant Ndebele (under 20% of the population), it was seen as one of Africa's comparatively stable countries, achieving a high literacy rate and economic growth. Some of its white minority left for South Africa, but many remained, accepting their loss of power. From the late 1990s, President Mugabe's rule was increasingly autocratic and arbitrary. His government seized white-owned land for redistribution without providing for maintaining production (thus bringing severe food shortages to a former agricultural exporter), rigged elections (most recently in 2002 and 2005), suppressed all political opposition, and (at considerable cost) intervened in the war in Congo. In 2005 the government suddenly began an urban 'clean-up' that destroyed the homes and businesses of 700,000 people (more than 5% of the total population), most of them poor. By 2006 the inflation rate reached 1,000%, unemployment was well over 50%, and at least half a million people had left the country. The 'brain drain' of educated emigrants, no longer largely white, made economic recovery unlikely in the short term.

In *Mozambique*, the leftist regime installed in 1975 was attacked by Renamo (Resistencia Nacional Moçambicana), a movement organized by Rhodesia and later backed by South Africa. The railway line and oil pipeline from Beira were vital links for Zimbabwe, and it sent a large force into Mozambique to guard them against the Renamo guerrillas, who for several years were in control of more than half of the country.

By 1992, Mozambique had suffered 16 years of civil war and a million deaths, many from starvation; droughts and ill-chosen economic policies had compounded the effects of the war. International mediation then led to the signing of a peace pact. A UN force was sent in; it withdrew in 1995, after an election had been held. Renamo, defeated in the election, was induced to accept the outcome; the leftists retained power, but modified their doctrinaire Marxism.

33 Angola and Namibia

The granting of independence to Angola in 1975, after 14 years of fighting, led to the withdrawal of 50,000 Portuguese soldiers and the flight of 300,000 Portuguese civilians. Three rival movements were left to fight a new guerrilla war against each other: in the north, the FNLA (Frente Nacional de Liberação de Angola), based on the Bakongo tribes; in the south, UNITA (União Nacional para a Independencia Total de Angola), based on the Ovimbundu, who form a third of Angola's population of 12 million; and the Marxist MPLA (Movimento Popular de Liberação de Angola), which held the capital, Luanda. A fourth group, FLEC (Front pour la Libération de l'Enclave de Cabinda), was trying to establish a separate state in Cabinda, the Angolan enclave north of the Congo river mouth, whose oilfields had begun production in the 1960s.

With the help of a South African force, UNITA nearly succeeded in capturing Luanda; the FNLA had the use of bases in Zaire; both, at first, had American support. But by 1976 the Cuban soldiers who were flown in by Soviet aircraft *(71)* had saved Luanda for the MPLA and enabled it to take over the north and west of the country. During the early 1980s the FNLA and FLEC gave up the struggle. UNITA fought on, and the MPLA government never mastered the south-east.

The struggle in Angola became linked with one in Namibia, the territory formerly called South-West Africa, which South Africa had taken over from Germany after the 1914–18 war. Much of Namibia is desert; it has only about 2 million inhabitants, half of whom are Ovambo, living in its far north; its most notable resources are the uranium mined at Rössing and the diamonds found south of Lüderitz. In the 1980s South African troops were sent to northern Namibia – including the 'Caprivi strip', named after the German minister who got it included in his country's former colony – to fight the guerrilla forces of SWAPO (South-West African People's Organization). These guerrillas operated from southernmost Angola, and the South African forces that were trying to suppress them repeatedly occupied areas there.

By the late 1980s Angola's MPLA regime was getting less Soviet support, and getting most of its revenues from American oil companies. In 1988, after long US mediation, Angola, Cuba and South Africa signed agreements under which the Cuban soldiers were to leave Angola by 1991, while in Namibia a long-delayed United Nations plan for the transition to independence would go ahead. South Africa withdrew its forces from Namibia, a UN peacekeeping force took over, and an election was held. In 1990, Namibia became independent. In its new legislature, an elected government based on SWAPO faced an elected opposition; the white minority (about 6% of the population) did not leave. In 1993, South Africa agreed to hand over its enclave at Walvis Bay to Namibia.

In 1991, Cuba withdrew the last of its forces from Angola (in all, it had sent 50,000 men),

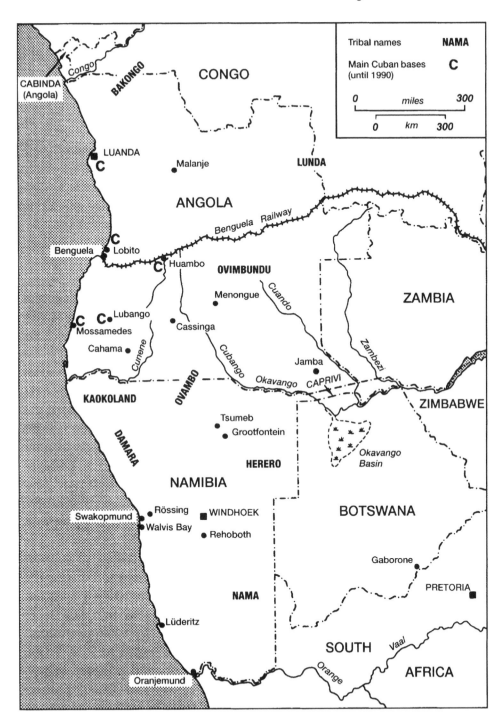

Tribal names   **NAMA**

Main Cuban bases   **C**
(until 1990)

0   *miles*   *300*

0   *km*   *300*

CABINDA
(Angola)

Congo

BAKONGO

CONGO

LUANDA
C

● Malanje

LUNDA

ANGOLA

Benguela Railway

Benguela
C
● Lobito

C ● Huambo

OVIMBUNDU

● Menongue

Cuando

ZAMBIA

C **C** ● Lubango
● Mossamedes

● Cassinga

Cubango

● Cahama
Cunene

KAOKOLAND

OVAMBO

Okavango ● Jamba

Zambezi

CAPRIVI

ZIMBABWE

DAMARA

● Tsumeb
● Grootfontein

● Okavango
Basin

HERERO

NAMIBIA

● Rössing   ■ WINDHOEK

BOTSWANA

Swakopmund
● Walvis Bay   ● Rehoboth

● Gaborone

PRETORIA ■

NAMA

● Lüderitz

SOUTH   *Vaal*

AFRICA

Oranjemund   *Orange*

and a peace pact was signed. In 1992 an election was held under UN supervision; the MPLA won. However, attempts to merge the two rival armies failed, and when UNITA found that it would not be given a large share of power it renewed the war – after only 18 months of peace.

By 1993, UNITA held two-thirds of the whole country, having captured Huambo in central Angola and important diamond-mining areas in the north (Angola, when at peace, is the world's fifth-biggest diamond producer, with 11% of all known reserves). But it then began to lose ground; it had lost South Africa's support – whereas the MPLA had started to hire South African mercenaries. A new peace pact, and the arrival in 1995 of a new UN force, gave Angola a second chance to emerge from a prolonged conflict which had cost more than 400,000 lives. Fighting resumed when the leader of UNITA refused to join a new government, but ended after government forces killed him in 2002; UNITA then disarmed, re-forming as an opposition political party. Democratic elections have yet to occur.

34 Republic of South Africa

In 1994, by an impressively peaceful transition, South Africa's long years of rule by its white minority were ended. The impact of this dramatic change was felt throughout Africa. It was the culmination of the process by which, since the 1950s, virtually the whole of the continent had emerged from a period of colonial or white-minority rule *(30, 31)*.

South Africa's population of 47 million includes more than 4 million whites. A majority of them are 'Afrikaners' (mostly of Dutch origin, speaking Afrikaans, a variant of Dutch), the rest being English-speakers and mostly of British origin. There are 37 million blacks, about a million descendants of immigrants from India, and 4 million 'Coloureds', of mixed origin. (In a country which in recent years was notorious for strict enforcement of race segregation, previous interbreeding had created a mixed-race community not much smaller than the white one.)

The first Dutch colony at Cape Town was established in 1652. Britain annexed the Cape of Good Hope in 1814. Thereafter, South Africa's history comprised two struggles for mastery: between British and 'Boers', and between whites and blacks. Discoveries of minerals, notably the gold of the Rand (Witwatersrand) near Johannesburg in the Transvaal, raised the value of the prize. In the late 1940s the Afrikaners emerged as victors in both contests – for a time.

In 1961, South Africa became a republic and left the Commonwealth. (It rejoined in 1994; Commonwealth actions had helped to ease its transformation.) In the 1960s Britain gave independence to three protectorates which South Africa's white rulers had once hoped to take over, Basutoland (now Lesotho), Bechuanaland (now Botswana) and Swaziland; but they remained economically dependent on the RSA.

From the 1950s onwards, a system intended to perpetuate white domination and segregate people by skin colour (*apartheid*) was continually tightened. An 'abolition' of the RSA's black majority was planned. Since 1913, blacks had been barred from owning land except in 'native reserves' (this, by forcing blacks to become tenants or employees, helped to provide cheap labour for white-owned farms and mines). The reserves were converted into 'homelands' for tribal groups, which were allowed a degree of self-government and encouraged to claim their 'independence'. When a tribe's homeland became nominally independent, all members of that tribe could be treated as 'foreign' – even though most of them lived in the 'white areas'. Thus, South Africa was to become a mainly white nation that happened to employ millions of foreign black 'immigrants' – who would be there on sufferance, with no rights.

By 1982, Bophuthatswana, Ciskei Transkei and Venda were 'independent'. And 3.5 million blacks had been deported from the 'white areas' to the homelands, where there was no employment or land for them. (But a majority of blacks still lived – mostly in segregated 'townships' – in the 'white areas', where they outnumbered the whites.)

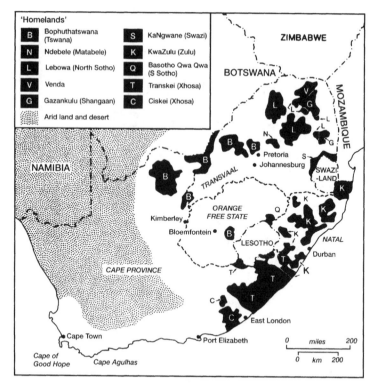

'Homelands'

B	Bophuthatswana (Tswana)	S	KaNgwane (Swazi)
N	Ndebele (Matabele)	K	KwaZulu (Zulu)
L	Lebowa (North Sotho)	Q	Basotho Qwa Qwa (S Sotho)
V	Venda	T	Transkei (Xhosa)
G	Gazankulu (Shangaan)	C	Ciskei (Xhosa)
	Arid land and desert		

In 1983 a new constitution created houses of parliament for the hitherto unrepresented Coloureds and Indians – while real power was kept in white hands. Most Coloureds and Indians, refusing to be used as tools against the black majority, boycotted the elections. Some joined blacks in protesting against the new constitution; this led to wider unrest. By 1986 the government had to impose a nationwide state of emergency.

The RSA had lost, by 1980, the sheltering northern screen that Portugal's colonies and white-ruled Southern Rhodesia had provided *(31, 32)*. Britain, formerly South Africa's chief trading partner, imposed some economic sanctions, and was urged by other Commonwealth states to impose more; in 1986 the United States also imposed sanctions. Big international firms abandoned their activity in the RSA; banks withheld credit; confidence cracked, and the economy went into recession.

In 1989 a new president, Frederik de Klerk, took a new course. By 1993 he had ended the state of emergency, legitimized the African National Congress (ANC) and other previously banned parties, freed Nelson Mandela and other long-imprisoned ANC leaders, and held talks with them about drastic reforms. In 1994, for the first time, all races voted on equal terms.

Mandela became South Africa's first black president. In the new parliament the ANC was dominant, but the National party (the formerly all-white and mainly Afrikaner party which had held power from 1948 to 1994) had a share in government until it left the coalition in 1996. The country's map was redrawn, the homelands being absorbed by the nine new provinces. The ANC controlled seven of the provincial governments; the Western Cape was run by the National party and KwaZulu-Natal by Inkatha, a party based on the biggest tribal group, the 10 million Zulus.

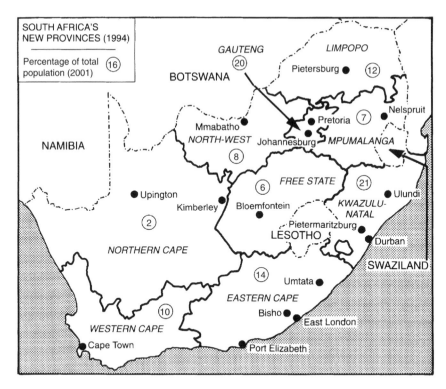

SOUTH AFRICA'S NEW PROVINCES (1994)

Percentage of total population (2001) (16)

Inkatha distrusted the ANC–National combination and in 1995 withdrew from the constituent assembly. It demanded more self-government for KwaZulu-Natal, and at times seemed to contemplate a secession. Violent clashes between Inkatha's supporters and the ANC's – in which more than 10,000 people had died – continued, although there were fewer of them after 1994 (and it was revealed that some of the earlier clashes had been fostered by officials in the former white-run administration).

Inkatha's traditionalist leaders were, like many whites, suspicious about the ANC's links with South Africa's Communist Party. But the sight of communism collapsing in eastern Europe and the Soviet Union, and of 'African socialism' causing economic disasters in various African states, seemed to have affected the ANC leaders before they came to power. They shelved their plans for massive state takeovers and sought to encourage foreign investment.

Although the RSA, thanks to the peacefulness of its transformation, remained a much richer country than its neighbours, its economy needed to resume rapid growth if the new government was to meet the blacks' expectations of a better life. At the same time, South Africa's leading export was in decline: between 1980 and 2004 its share of world gold production fell from 56% to 14% as extraction from its deep mines became more difficult. Removing the distortions of *apartheid* required economic diversification and massive investment in housing, education and health care. To the surprise of those who feared the worst in 1994, a decade later the economy was growing, inflation was down, and South Africa remained firmly democratic. High unemployment remained, however, and more than 5 million South Africans were infected with HIV *(30)* – their plight worsened by the failure of some South African politicians to acknowledge the link between HIV and AIDS, and to support effective treatment programmes.

35 Sudan and the Horn of Africa

Sudan, formerly an Anglo-Egyptian condominium (in practice ruled by the British), became independent in 1956. The dominance of its Arabized Muslim north has long been resisted by the south's black Africans, who prefer indigenous religions or Christianity, and the government's support of terrorism has not been popular abroad. In 1983, Islamic laws were imposed on the south; a new revolt began, and by 1990 the rebels held most of the south. In 1991 they split, on tribal lines; this intertribal conflict continued until 1995. Three years of negotiation led to a peace agreement between north and south in 2005 that specified a six-year waiting period with autonomy for the south, to be followed by a referendum on southern independence. By the time of the agreement war and famine had caused 2 million deaths. Distribution of oil revenues will be a factor in a final peace – Sudan began to export oil in 1999, and most of its deposits are in the south.

Separate conflicts continued in the western region of Darfur and, on a smaller scale, in the east near the Red Sea. In Darfur, black African groups rebelled against the Arab-dominated government in 2003. Both are Muslim, but the nomadic Arabs and black farmers compete for land and water. More than 200,000 people died as the government suppressed the rebellion and supported Arab militia groups; millions more fled from their homes, some into Chad. A 7,000-strong African Union (AU) peacekeeping force was sent but it was poorly equipped and too small to patrol an area the size of France; the AU has requested NATO or UN support.

In 1996, encouraged by the success of the southerners, new rebel groups backed by Eritrea and Ethiopia began to launch attacks in north-east Sudan – activity of particular concern to the government because it threatens oil exports.

Ethiopia (formerly Abyssinia) was conquered by fascist Italy in 1936 but liberated by Britain in 1941. It has its own distinctive Christian (Coptic) church; but many of its inhabitants, outside the central Amharic-speaking region, are Muslims. (Most of the old community of Ethiopian Jews, called Falashas, were flown to Israel in 1984 and 1991.) A coup in 1974 installed a Soviet-backed military junta. The emperor, Haile Selassie (formerly Ras Tafari), was imprisoned and later killed.

Eritrea, a former Italian colony, was federated with Ethiopia in 1952. Ethiopia imposed direct rule there in 1962, igniting resistance. By the late 1970s large areas were held by the rebels, and another rebellion had begun in the neighbouring Ethiopian province of Tigre.

Somalia became independent in 1960, uniting the former Italian Somaliland with the former British Somaliland (the area around Berbera). Nationalist stirrings among the Somalis in the Ogaden region of Ethiopia, and in *Djibouti*, then a French territory, led to frontier disputes and clashes. Djibouti became independent in 1977, its Issa (Somali) and Afar leaders agreeing to share power. France retained a base there.

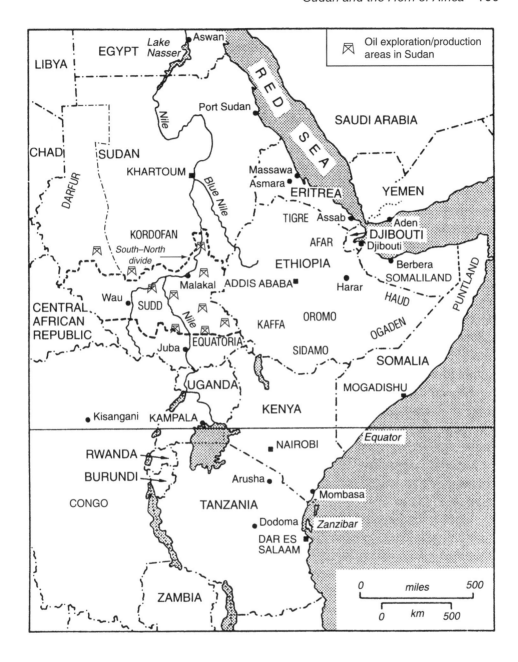

Somalia's quarrels with Ethiopia induced the latter to let the Soviet Union install bases on its coast *(42)*. In 1977 a Somali revolt in the Ogaden, supported by Somalia's forces, developed into a full-scale war, which the Ethiopians won because the USSR flew in 20,000 Cuban soldiers to fight for them *(71)*. From then on, the USSR provided arms and 'advisers' for Ethiopian forces fighting in Eritrea and Tigre as well as on the Somalia frontier, while Somalia expelled the Russians and invited the Americans to take over the former Soviet bases; they duly did so.

In 1988, Ethiopia and Somalia disengaged their forces and restored diplomatic relations. Then rebellions began in Somalia, starting in the north. In 1991 the ex-British area was declared an independent 'Somaliland', with its capital at Hargeisa. Since then, though unrecognized by other countries, Somaliland has been the most stable part of the 'failed state' – Somalia has not had a central government since the military ruler fled in 1991. A chaotic struggle between clans and warlords ensued, and a famine in which 400,000 people died. In 1992, American troops went in to protect the supply of relief food; in 1993 a UN force was installed; the famine was overcome, but all efforts to stop the fighting failed. The last American soldiers left in 1994, the last UN ones in 1995. The north-east corner of the country broke away as 'Puntland' in 1998, not claiming full independence but providing a buffer between Somaliland and any future attempt at central governance.

Famine had become a recurrent threat in a region where relief efforts were often obstructed, by wars or by rulers unconcerned about their starving people (Ethiopia's Marxist rulers used hunger and long-distance forced migration as means of subduing unrest). And the Somalis' persistent fighting in the 1990s showed the limitations of any intervention aimed at stopping an internal war. There were no religious, and few linguistic or ideological, divisions among the 9 million Somalis; nor were foreign powers trying to divide them. Yet fighting continued in 2006 between Islamists (who gained control of the capital) and an Ethiopian-supported 'transitional' government.

In the late 1980s the USSR withdrew the Cubans from Ethiopia, whose army then lost ground. By 1991 it had been smashed. A victorious coalition, based on the Tigrayan rebel movement but also representing other groups – including the Oromo (Galla), who form 40% of Ethiopia's population – took power. It began a decentralizing process, ending the old Amharic domination and redrawing provincial boundaries along ethnic lines.

Eritrea was left free to secede, and after a 1993 referendum it became independent. A dispute over several small, arid, sparsely populated areas (without valuable natural resources) sparked a short war in the late 1990s. After a ceasefire in 2000, both sides initially accepted a boundary drawn by an independent commission, but in 2003 Ethiopia objected when the town of Badme was awarded to Eritrea, and a UN peacekeeping force remains.

The recent conflicts in eastern Africa (see also *36*) were interlinked in many ways. Refugees flooded across most of the region's frontiers. Each state, at times, supported rebels in one or more of the other states. Rebel forces crossed and recrossed frontiers. There were also many minor frontier clashes between states' armies. But the only full-scale wars between states were the Ethiopia–Somalia and Uganda–Tanzania wars in the 1970s. Most African countries have boundaries drawn by European colonial powers with little regard for historic African states or ethnic regions, so ethnic groups are grouped together or split by national boundaries. Thus, decentralization of power and secession – so far pursued successfully only by Eritrea, but demanded in Sudan and Somalia – are more prevalent than international conflict.

36 East Africa

Kenya, *Uganda* and *Tanzania* (Tanganyika and Zanzibar were united in 1964 to form Tanzania) had been British-ruled before they came to independence in the 1960s. By 1977 their East African Community had broken up. Kenya and Tanzania were then on such bad terms that the frontier between them was closed until 1983. Later, relations improved, but the close co-operation of the 1960s was not restored.

In Uganda, President Obote's arbitrary rule, including his abolition of tribal monarchies, caused unrest, especially among the Baganda, the people of the old kingdom of Buganda. The army commander, General Amin, seized power in 1971. He imposed a reign of terror, and expelled the whole Asian community; and in 1976 he collaborated with Arab and neo-Nazi German hijackers who brought a planeload of hostages, including many Israelis, to Entebbe airport. (An Israeli airborne commando rescued the hostages.) In 1978, Amin invaded Tanzania. The counterattacking Tanzanians quickly overran Uganda; troops were sent by Libya *(39)* to help Amin run away, and so he did. With Tanzanian backing, Obote regained power; but his army proved as brutal as Amin's, provoking a resistance movement whose forces captured Kampala in 1986. They installed a more able and less oppressive government that took some steps towards democracy and has been among the most successful in Africa in combating the spread of AIDS. Unfortunately, the change of regime in 1986 was also followed by a rebellion in the north, where the Lord's Resistance Army (LRA) has murdered and kidnapped thousands of people (mostly children). The rural population in several districts fled their homes for camps that themselves have come under attack by the LRA, most of whose 'soldiers' are enslaved children.

Rwanda and *Burundi*, small but populous highland states, became independent in 1962 after periods of German and then Belgian rule. In each of them, a Tutsi (Watusi) minority – 15% of the population – had for centuries dominated the Hutu (Bahutu) majority.

In Rwanda, an upheaval in 1959–60 had broken the Tutsis' domination, and many had fled into Uganda. These Tutsi exiles made several attempts to invade Rwanda. In 1990 they came in greater strength, and the government called in French and Belgian troops. When those troops withdrew, the invading Tutsi army resumed its attacks.

By 1994 it was again threatening the capital, Kigali, and in July it captured it; 800,000 Tutsi exiles then came home from Uganda; 2 million Hutus fled, mostly into Congo (Zaire). France sent to Rwanda a force which from June to August held a zone in which some Hutus found temporary refuge. Three months before Kigali fell, there had been a massacre of about 700,000 Tutsis and a smaller number of Hutus. Though the Hutu forces were armed with little more than machetes and there was a small UN force in place, neither the UN nor any other outside power intervened to stop the genocide while it was occurring. The murder was largely the work of Rwanda's Hutu troops and militia. Many thousands of these

were soon killed or captured by the victorious Tutsi invaders, but others escaped into Congo. There they took control of camps holding 1.2 million Hutu refugees.

In 1996 a rebellion in Congo's eastern region of Kivu *(32)* was provoked by attempts to expel its local Tutsis. Rebel Congolese Tutsis, called Banyamulenge, helped by Tutsi forces from Rwanda, took Bukavu and Goma, and drove the Hutu militia away from the refugee camps. Hundreds of thousands of refugees fled back from Congo into Rwanda, whose Tutsi rulers handled them roughly, or into Burundi, whose Tutsi army killed many as soon as they arrived. Tanzania's army then drove another 350,000 Hutu refugees back into Rwanda. In 1997, Rwanda (and Uganda) supported the overthrow of the Congolese government *(32)*, expecting that the new regime would help defeat the Hutu militias. When it did not, they invaded eastern Congo again in 1998; a peace agreement was eventually signed four years later, in which Rwanda agreed to withdraw from Congo and Congo promised to help suppress the militias.

In Burundi, Tutsi rule continued, and until 1993 many Hutu risings were bloodily suppressed (300,000 Hutus were killed in 1972). Then an election was held, producing a

Hutu president. Tutsi soldiers murdered him and, in an attempted coup, killed 200,000 Hutus. A semblance of coalition government survived, but Tutsis still controlled the army. In 1996, Tutsi military rule was reimposed; more massacres of Hutus ensued. Fighting and talks both continued through 2003, when an agreement with Hutu rebels ended the war; voters supported a power-sharing constitution in a 2005 referendum.

37 Nigeria and Guinea Coast

Around the Gulf of Guinea, five European nations' colonial rivalries left a patchwork of frontiers *(27)*. Before 1957 the only sovereign state was Liberia, where American-sponsored settlement of freed slaves had begun in the 1820s (the British did the same in Sierra Leone without allowing independence); but by 1975 all the colonies had become independent.

Germany's two pre–1914 colonies, Togo and Cameroon, were later each divided into British and French territories. In 1960 the French portions became independent states and, after UN-supervised plebiscites, British Togoland joined Ghana. The northern part of British Cameroon joined Nigeria and the southern part joined ex-French Cameroon, but a border dispute between Nigeria and Cameroon lingered until its settlement in 2006.

Britain gave independence to Ghana in 1957, to Nigeria in 1960, to Sierra Leone in 1961 and to the Gambia in 1965. France gave it to Guinea in 1958, to its other dependencies in the region in 1960; Spain gave it to Equatorial Guinea in 1968; Portugal to Guinea-Bissau in 1974, to the Cape Verde Islands and to São Tomé and Principe in 1975.

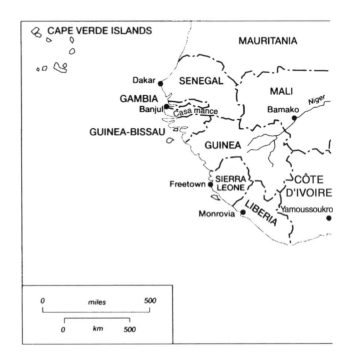

Before independence, Ghana had been the Gold Coast; Mali had been (French) Soudan; Guinea, Guinea-Bissau, and Equatorial Guinea had been French, Portuguese and Spanish Guinea. In 1975, Dahomey renamed itself Benin. In 1983, Upper Volta became Burkina Faso.

'West Africa' is not an exact term, but is often used to mean Nigeria, Niger and the fourteen states to the west of them. In 1975 these 16 set up the Economic Community of West African States; but ECOWAS remained a very loose grouping. 'Senegambia', a confederation formed by Senegal and the Gambia in 1981, broke up in 1989.

Since independence, political instability has been rife in a region whose boundaries were drawn by colonial powers without regard to ethnic and religious divisions. By 1999, Senegal was the only one of the 16 West African states that had not suffered at least one military coup. In several countries, soldiers who had seized power later staged 'elections' and became 'civilian' presidents. By the 1990s there was growing impatience with the autocrats' rule, and some of them had to permit contested elections. Most notably, in Benin in 1991, the sitting president ran for re-election, lost, and stood down peacefully – the first such transfer of power in the region. Ghana and Cape Verde have moved in the same direction. But democracy and stability still elude the countries that are home to four-fifths of West Africans.

In *Liberia* a confused civil war that began in 1989 had reduced the country to chaos by 1990, when a mainly Nigerian peacekeeping force, created by ECOWAS, was sent in. By 1993 there seemed to be prospects of peace, with rival leaders agreeing to form an interim council; in 1996 large-scale fighting was renewed; then a new truce brought an agreement to hold a presidential election in 1997. This was won by rebel leader Charles Taylor, known for his use of child soldiers and mutilation of enemies. He was soon accused of supporting the equally brutal rebellion in neighbouring Sierra Leone by providing weapons in exchange

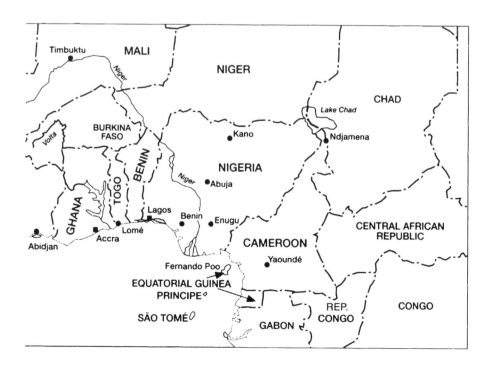

for diamonds. A new rebellion forced Taylor to flee to asylum in Nigeria in 2003 (he was later turned over to an international war crimes tribunal); intervention by Nigeria and the UN followed. A new election was held in 2005.

Meanwhile a similarly devastating civil war had begun in 1991 in neighbouring *Sierra Leone*, then under military rule. During a brief peace in 1996–7 Sierra Leone was able to hold an election and restore civilian government. The rebel Revolutionary United Front then mounted a coup; they fought for control of Freetown with a Nigerian-led ECOWAS force. Under a peace agreement the rebels disarmed, starting in 2001.

A religious dividing line runs across West Africa, from Senegal to northern Nigeria. The north is mainly Muslim, the south mainly adheres to Christianity and indigenous religions. The division is particularly significant in Nigeria, which has been riven by internal antagonisms – mostly recently following the introduction if *sharia* (Islamic law) in northern states – and Ivory Coast (Côte d'Ivoire), which has been split by a northern rebellion since 2000.

Nigeria, Africa's most populous country (its 130 million inhabitants speak hundreds of different languages), became independent as a federation of three regions. Its huge northern region was predominantly Muslim; in its western region the Yoruba people were dominant, in its eastern region the Igbo (Ibo). After two military coups in 1966, some Igbo leaders proclaimed at Enugu in 1967 a secessionist state called Biafra. They were not subdued until 1970. This civil war brought mass starvation in the Igbo area and bitter dissension among other African governments, several of which supported Biafra.

Nigeria's military rulers reshaped it as a federation of 19 states (since increased to 36); Abuja was chosen to succeed Lagos as the federal capital. From 1979 to 1983 a new attempt was made to maintain an elected government. But the economic boom that Nigeria had enjoyed in the 1970s ended as world oil prices fell *(3)*; the 2 million immigrants (mainly Ghanaians) whom the oil boom had drawn in were brusquely ordered to leave; military rule was reimposed. Increasingly, the military regime brought open domination by Muslim northerners. In 1986, Nigeria joined the Islamic Conference Organization *(28)*.

Corruption and wildly extravagant spending left Nigeria with no long-term benefit from its boom years. Between 1983 and 1996 its national income per head was halved. An election held in 1993 was annulled; the generals imposed military rule again. In 1995 they used great brutality in suppressing unrest among the 500,000 Ogoni in the Niger Delta oilfield area in the south-east. The Commonwealth suspended Nigeria's membership from 1995 to 1998 *(9)*; the European Union imposed an arms embargo and other sanctions.

A new constitution restored civilian rule in 1999 and promised democracy. Negotiations with creditor countries in 2005 helped reduce Nigeria's debt burden, and high oil revenues were being used to pay off the remaining debt. The high price of oil also increased tensions among the peoples of the Delta region, the government and oil companies. Delta residents protested that they remained poor and suffered from environmental damage; they attacked oil company facilities, forcing a reduction in exports.

38 Ex-French Africa

France's former territories in north and west Africa *(27)* formed a continuous area stretching from the Mediterranean to the Congo river. In this area there are now 17 sovereign states. In the north, Morocco, Algeria and Tunisia are Arab countries *(29, 39)*. The other 14 states, formerly grouped as French West Africa and French Equatorial Africa, have a combined population barely equal to Nigeria's; the Sahara desert covers most of Mauritania and much of Mali, Niger and Chad.

Of these 14, Guinea became independent in 1958, the others in 1960. The formerly French Republic of Congo has often been called Congo-Brazzaville to distinguish it from the Democratic Republic of Congo (Congo-Kinshasa), formerly the Belgian Congo and later Zaire. The Central African Republic (formerly Oubangui-Chari) was declared an 'empire' in 1976 by its military ruler, Colonel Bokassa, but he was ousted in 1979 and the CAR resumed its claim to be a republic. Mauritania's iron ore, Guinea's bauxite, Gabon's oil and Niger's uranium are notable resources in an area otherwise poorly endowed and much afflicted by drought in the 'Sahel' belt along the Sahara desert's southern edge.

After independence, French economic aid remained important for most of the 14 states; and the currency shared by 12 of them, the CFA (African Financial Community) franc, was tied to the French franc (now the euro). France kept bases in Cameroon, the CAR, Chad, Gabon, Ivory Coast and Senegal. In the CAR and elsewhere, the presence of small French forces sometimes helped to preserve, or to change, a government.

Before the breakup of the Soviet Union, Soviet influence was at various times strong in Benin, Congo and Guinea. In the 1970s and 1980s, however, the most disturbing factor in the region was the expansionist ambition of Libya's military ruler, Colonel Muammar Qaddafi. Libyan diplomats were caught smuggling explosives into Senegal, and were involved in coup plots in Benin and elsewhere. The 1983 coup in Burkina Faso installed a regime which accepted Libyan subsidies.

The scene of the most open Libyan intervention was *Chad*. In 1973 the Libyans occupied Chad's 'Aozou strip', hoping to find uranium deposits like the rich ones around Arlit in Niger *(3)*. In 1980 they sent troops to Ndjamena, Chad's capital, and in 1981 they proclaimed the union of Chad with Libya. The Organization of African Unity got the Libyans to withdraw; but in 1982 they established at Bardai, near Aozou, a 'government' of dissident Chadians and, using this as a cover, they sent their forces into Chad's northern desert areas. A small French force sent to Chad for a few months in 1983–4 deterred the Libyans from attacking Ndjamena, and in 1987 Chad's own forces drove them back to Aozou. Mediation by the OAU brought a ceasefire. The International Court at The Hague was asked for a ruling on the Aozou strip; in 1994 it ruled in favour of Chad. By then, fighting among

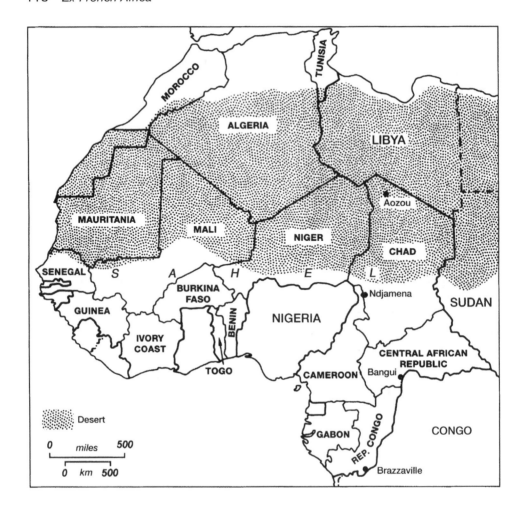

Chadians had brought a new government to power; its relations with Libya were less hostile, but it was not regarded as a Libyan puppet.

Chad began exporting oil through a pipeline to a port in Cameroon in 2003. The government promised to devote most of the revenue to health, education and development, but later sought to divert more for military purposes.

War in Sudan brought a confusing mix of conflicts to Chad, starting in 2003. Refugees from Darfur in Sudan *(35)* escaped into Chad – as elsewhere in Africa, ethnic ties cross national borders – but were followed by Sudanese troops. Anti-government Chadian rebels advanced from Darfur to Ndjamena in 2006 but were defeated; Chad accused Sudan of supporting them and, in turn, supports the rebels in Darfur. France supports the government of Chad and maintains more than 1,000 troops in the capital.

Oil wealth did little to improve living standards or government elsewhere in the region. Gabon, sub-Saharan Africa's third-largest producer, managed to build up a large foreign debt and has made no effort to diversify its economy. Oil was discovered in neighbouring Equatorial Guinea in 1996, making its 500,000 citizens among the world's richest people – yet as late as 2006 they could expect to live only fifty years on average (compared to fifty-four in Gabon). In Gabon, the president has held office since 1967; in Equatorial Guinea, since 1979.

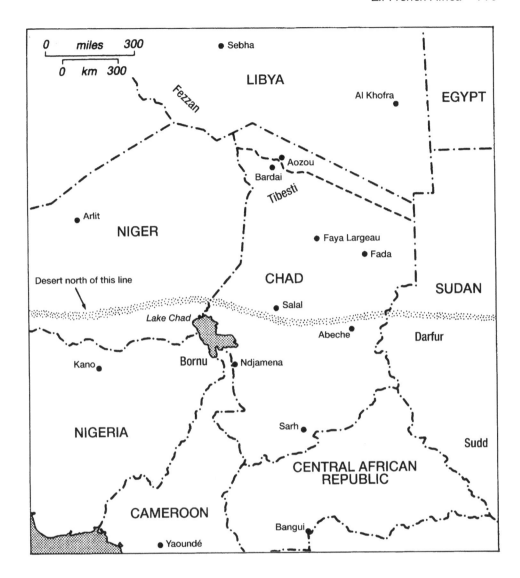

Congo-Kinshasa *(32)* and its small eastern neighbours, Rwanda and Burundi *(36)*, had been Belgian territories, not French. However, they joined the '*Francophonie*' grouping *(9)*; and French forces were sent to these countries several times – notably to Rwanda in 1994. That action led to charges that France had sided with the Hutus against the Tutsis in Rwanda's internal struggles.

39 North Africa

In the 'Maghreb' (the 'West' of the Arab world), independence was given in 1956 to two French protectorates, Morocco and Tunisia; but Algeria did not achieve it until 1962, after seven years of fighting. The turmoil this war caused in France was ended by President Charles de Gaulle when, defying and outmanoeuvring the powerful 'French Algeria' interests, he withdrew the army and recognized an independent Algeria. Since France first colonized it in the 1830s a million Europeans had settled in Algeria; nearly all left after 1962. A migration to France of about 3 million Algerians, Moroccans and Tunisians also developed.

Although these countries' main culture is Arabic – with French elements superimposed – the pre-Arab Berber language *(29)* survives in parts of Algeria and Morocco – notably in Kabylia, east of Algiers, where demands for the protection of Berber culture increased in the 1980s. Tamazight, the Berber language, was recognized as a 'national' language in 2001 in Algeria, but it was not granted full 'official language' status alongside Arabic. The situation is similar in Morocco, where a larger share of the population is Berber.

All three countries are Muslim (and Morocco's kings have long been religious leaders as well as rulers), but Islamic fundamentalism did not become a major force until the 1980s,

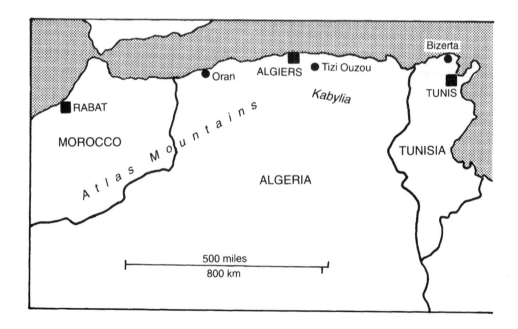

when it began to pose a serious challenge to Algeria's military-dominated regime. The 1992 election in Algeria was annulled when the Islamic Front was seen to have won it. A bloody conflict, akin to civil war, developed. Foreigners were targeted by some of the Algerian fundamentalist groups; many were killed, all were threatened with death unless they left the country, and most of them duly left. In 1996 the French government urged the remaining 8,000 French in Algeria to leave. Violence eased in the late 1990s, and a presidential election held in 2004 was comparatively free, by regional standards.

In the 1960s, Algeria and Morocco had fought over disputed frontiers, and after 1975 Algeria became the main backer of the guerrillas who were opposing Morocco in Western Sahara *(40)*. Diplomatic relations, broken off in 1976, were not resumed until 1988. By then Algeria and Morocco were on much better terms. In 1989 they joined with Libya, Mauritania and Tunisia in forming the Arab Maghreb Union. One of the AMU's chief aims was to achieve unity in trade negotiations with the European Union, the market for most of the North African states' exports (mainly oil and gas from Algeria and Libya, fruit and wine from Morocco and Tunisia, iron ore from Mauritania).

Libya became independent in 1951; it had been taken from Turkey by Italy in 1911 and fought over by the British, the Germans and the Italians during the 1939–45 war. It is mostly desert, but was spectacularly enriched in the 1960s by newly found oilfields. In 1969 army officers led by Colonel Muammar Qaddafi seized power and embarked on ambitious foreign policies, using the huge oil revenues, terrorist tactics and a special brand of 'Islamic' slogans (not those of the fundamentalists; Qaddafi denounced them as 'imperialist lackeys', to be hunted down and killed).

Libya sent forces to fight, unsuccessfully, in Uganda and Chad *(36, 38)*. It financed and backed attempts to overthrow the governments of several Arab and African states; its assassins, armed and directed from 'diplomatic missions', also operated in European countries and America. In 1984, after a series of murders in Britain, machine-gun fire from the Libyan mission in London killed a policewoman. When gunmen were caught (as in

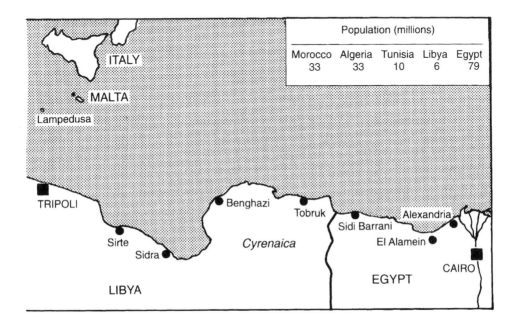

Population (millions)				
Morocco	Algeria	Tunisia	Libya	Egypt
33	33	10	6	79

Germany in 1983, in Italy in 1986) Libya took foreigners hostage to enforce their release. Qaddafi armed and financed German, Irish and other terrorist groups; in 1989 he openly admitted this.

Qaddafi's sea claims also caused tensions; in 1980 his warships forced Malta to stop exploring for oil 60 miles from its coast; in 1981 and 1986 his aircraft attacked US planes over the Gulf of Sirte, which Libya claimed as 'internal waters'. The United States held Libya responsible for the killing of soldiers in Berlin in 1986; in reprisal, US aircraft bombed Tripoli and Benghazi. Other cases in which suspicions were directed at Libya included the 1984 laying of mines which damaged seventeen ships in the Red Sea, and the destruction by bombs of a US airliner over Lockerbie in Scotland in 1988 and of a French airliner over Niger in 1989. United Nations sanctions were imposed on Libya in 1992–3 after it refused to hand over the agents believed to have planted the Lockerbie bomb.

Later in the 1990s, Qaddafi sought to normalize relations with Europe and the US by atoning for past sponsorship of terrorism. In 1999 he turned over two suspects for trial in the Lockerbie case, one of whom was convicted and imprisoned. Libya later paid compensation for the Lockerbie, Berlin and Niger attacks. Qaddafi even softened towards Israel, favouring a peaceful resolution to Arab–Israeli conflicts. But he remained a dictator with a poor tolerance for dissent; Libya's election as chair of the UN Commission on Human Rights *(7)* was regarded as a scandal. In 2003, UN sanctions were lifted and Libya formally abandoned its pursuit of nuclear weapons. American oil companies were allowed back into the country in 2005, and the US restored diplomatic relations in 2006.

40 Morocco and Western Sahara

Before Morocco became independent in 1956 most of it was under French control, but Spain held zones in the north and the south, and Tangier was an international zone. Spain retained an enclave at Ifni until 1969, and it still holds Ceuta and Melilla *(25)*. As the only remaining European territories in Africa, they attract migrants from North and West Africa who seek admittance to the EU; construction of border barriers has driven some to go by way of the Canary Islands instead.

Morocco's monarchy was slightly liberalized in the late 1990s, and elections for a parliament were allowed in 2002.

In 1975, Spain withdrew from its Western Sahara territory – a thinly peopled, largely desert area, whose only notable resources were phosphates near Bou Craa. Morocco, which had old claims to the territory, sent in troops, while Mauritania took over part of the southern region (Rio de Oro). They met resistance from a guerrilla movement, backed by Algeria and based on the Tindouf area, called the Popular Liberation Front of Sakia el Hamra and Rio de Oro (Polisario).

In 1979, Mauritania withdrew its forces. Morocco then sent troops into the south, but found it hard to prevent raiding there by the guerrillas, who could get there from Tindouf by crossing Mauritania's desert areas. Between 1980 and 1989 the Moroccans built a chain of sand 'walls' as a protection for almost the whole territory.

In 1991, Morocco offered to hold a referendum, so that the Sahrawis (Saharans) could choose between independence and attachment to Morocco. But there was no agreement about who should be entitled to vote. Before 1975 many Sahrawis had been nomads, moving in and out of the territory. Now many were living in Polisario-controlled refugee camps around Tindouf; and many people had migrated into Western Sahara from Morocco. The United Nations arranged to send observers and a small UN force to supervise a referendum; but the unresolved dispute over voting rights prevented the referendum from being held.

By 1984 a majority of African states had recognized the Sahrawi government-in-exile proclaimed by the Polisario leaders. When Sahrawi delegates were allowed to attend meetings of the Organization of African Unity (OAU), Morocco withdrew in protest. By 1990, however, Polisario was no longer being given arms by Libya or – more importantly – by Algeria; its guerrillas remained based on Algerian territory, but they made few attempts to penetrate the 'walls'. The improvement in relations between the Arab governments of the Maghreb *(39)* had made Algeria less eager to fight a proxy war with Morocco through Polisario. Between 1993 and 1999 an effort was made to draw up a list of voters for a referendum, but agreement on who would be eligible was still lacking in 2006.

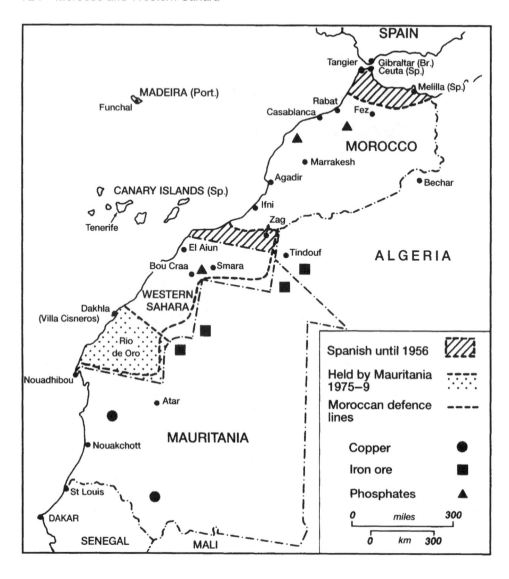

SPAIN

Tangier
Gibraltar (Br.)
Ceuta (Sp.)
Melilla (Sp.)

MADEIRA (Port.)
Funchal

Rabat
Casablanca
Fez

MOROCCO

Marrakesh

Agadir
Bechar

CANARY ISLANDS (Sp.)
Ifni

Tenerife
Zag

El Aiun
Tindouf

ALGERIA

Bou Craa Smara

WESTERN
SAHARA

Dakhla
(Villa Cisneros)

Rio
de Oro

Nouadhibou

Atar

MAURITANIA

Nouakchott

St Louis

DAKAR

SENEGAL
MALI

Spanish until 1956

Held by Mauritania
1975–9

Moroccan defence
lines

Copper

Iron ore

Phosphates

0   miles   300

0   km   300

41 Middle East and North African Oil

Oil production began in Iran (Persia) in 1912; in Iraq in the 1920s; in Saudi Arabia in 1939; in Kuwait in 1945. By the mid–1950s the Middle East was supplying three-quarters of Europe's needs. The Arab countries of North Africa then began to produce oil. By 1960 the Middle East and North African oil-exporting states – all of them Arab, except Iran – were producing 25% of world output; by 1970 they produced 40% of it. Their oil was plentiful, easily accessible (even in the offshore fields, which lie under relatively shallow water) and thus cheap to extract. Much of it came from desert areas, and it enriched formerly poor countries such as Saudi Arabia and Libya.

The oilfields were developed by American, British, Dutch and French companies. In 1951, Iran expropriated the British company working there and seized its refinery at Abadan. Thereafter the exporting states' governments steadily tightened their grip on the industry, increasing taxes, nationalizing companies, and creating their own state corporations.

The first pipelines built to carry Iraqi and Saudi oil to the Mediterranean were put out of action by the successive conflicts that involved Israel, Lebanon and Syria. Tanker traffic from Gulf ports to Europe through the Suez Canal was halted by the conflicts that closed the canal in 1956–7 and 1967–75 (43, 44). The longer 'Cape route' round Africa had to be used, huge supertankers were built for this purpose, and the deepening and widening of the canal in 1980 did not bring all the oil traffic back to it. Meanwhile new pipelines were built: one through Egypt and one through Israel, both from the Red Sea to the Mediterranean; one from Iraq across Turkey to the Mediterranean at Ceyhan; others carrying Saudi and Iraqi oil to Yanbu on the Red Sea.

Iraq deprived itself of the ability to export oil by tanker when it invaded Iran in 1980; traffic from its terminal near Basra was promptly blocked. Its invasion of Kuwait in 1990 led to the blocking of the pipelines carrying Iraqi oil across Turkey and Saudi Arabia. Iraq's exports remained completely blocked (except for a little oil sent out by road) until 1996, when it accepted the conditions that the United Nations had set for a strictly limited resumption (48). These rules remained in effect until war began again in 2003.

One reason for the strong international reaction to the invasion of Kuwait was that, of all known reserves of conventional oil (excluding oil sands [3]), about 10% are in Iraq, 10% in Kuwait and 25% in Saudi Arabia. If the Iraqis' annexation of Kuwait had been tamely accepted, they might have been emboldened to seize the adjacent Saudi oilfields, too; they would then have had control of 45% of the world's oil reserves.

For a long time the Soviet Union's proximity to the oil-rich Gulf region, with its vulnerably weak states and local conflicts, had caused recurrent anxiety in the west – particularly after the Soviet invasion of Afghanistan in 1979, which followed the withdrawal

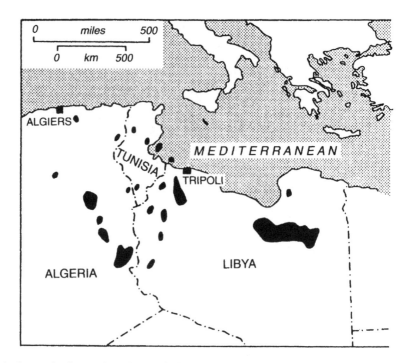

of the last British forces in the region *(42)* and the upheaval that ended Iran's role as a western-armed buffer against possible Soviet southward thrusts *(47)*. But Iraq's war with Iran and invasion of Kuwait, which posed more immediate threats to the flow of Gulf oil, led westerners to regard the USSR, in its last few years, as potentially helpful in preserving the region's stability.

In the early 2000s about one-third of world oil output was coming from the Middle East and North Africa. Saudi Arabia was much the biggest producer in OPEC *(3)*. Europe's formerly acute dependence on the Middle East for fuel had been somewhat reduced by

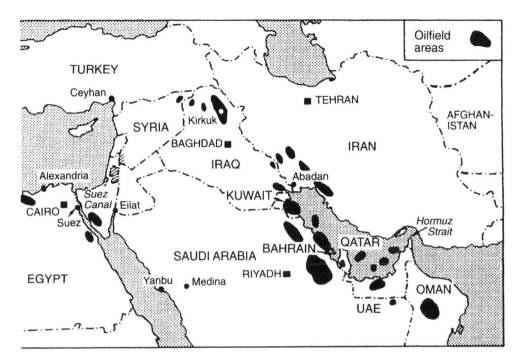

developments elsewhere – including the exploitation of North Sea oil and gas *(22)* and the piping of gas to Europe from Russia *(77)*. Japan and China, however, had become massive consumers of Gulf oil, which was also providing a quarter of American imports. The world had repeatedly seen how its whole economy could be drastically affected by disturbances in the region *(3)*; and there was good reason for continuing international concern about a region that contained two-thirds of all the reserves of oil that were then known. Rapid economic growth in oil-poor China and India promised to squeeze supplies further, despite more efficient use of energy in wealthier importing countries since the 1970s.

42 Suez and Indian Ocean

The Suez Canal was built in the 1860s by a French-based international company, by agreement with the rulers of Egypt and of the Turkish empire. In 1882 the British occupied Egypt, completing a strategic chain in which the main links were Gibraltar, Malta, the Suez Canal and Aden. The Suez route became Britain's imperial lifeline to its possessions in the east, which at one time included India, Burma, Malaya, Australia, New Zealand, much of East Africa and many of the Indian Ocean islands. Later, Britain became dependent for three-quarters of its oil supplies on tankers from the Middle East passing through the canal *(41)*; its own North Sea oilfields were not fully developed until the end of the 1970s.

Under a 1936 treaty with Egypt, British forces withdrew from most of the country but remained in the canal zone. In the 1939–45 war, when British and Commonwealth forces repelled German and Italian attempts to seize the canal and reach the Indian Ocean, the canal zone became a major British base. This base was evacuated, at Egypt's demand, a few months before the 1956 Suez conflict *(43)*.

Meanwhile the British relinquishment of empire had begun with the achievement of independence by India and Pakistan in 1947. 'East of Suez', the process had almost been completed by 1968, when Britain announced plans to remove its remaining forces from the small Gulf states and Singapore by 1971 *(47, 63)*. Among the Indian Ocean islands, Britain gave independence to the Maldives in 1965, to Mauritius in 1968 and to Seychelles in 1976 (when the islands of Aldabra, Desroches and Farquhar were transferred to Seychelles control).

With the era of British predominance in the Indian Ocean thus ended, the Suez Canal lost much of its strategic importance. However, the new situation in the region carried echoes of the nineteenth-century period when the British saw Russia's conquest of central Asia *(54)* as a threat to their Indian empire, and feared that Russia might reach the Indian Ocean by way of Iran (Persia). Tsarist Russia had been deterred from trying to take control of Afghanistan; the Soviet Union was not. When its attempts in the 1970s to manipulate a client regime there failed, it sent in an army which waged a ten-year war. A million Afghans were killed, and more than 5 million fled from their devastated country into Iran and Pakistan.

Continuing Afghan resistance forced the USSR to withdraw its army in 1989; but in the previous two decades it had established itself in the Indian Ocean as a naval power, acquiring bases for its fleet in Vietnam, at Aden in South Yemen, in Somalia, and later in Eritrea, then part of Ethiopia. Largely because of this new extension of Soviet naval power, the Americans took to maintaining a similar presence in the Indian Ocean. One of their moves was the building of an airfield on Diego Garcia, in the Chagos group – officially, since 1965, the British Indian Ocean Territory. This last-remaining British dependency in the

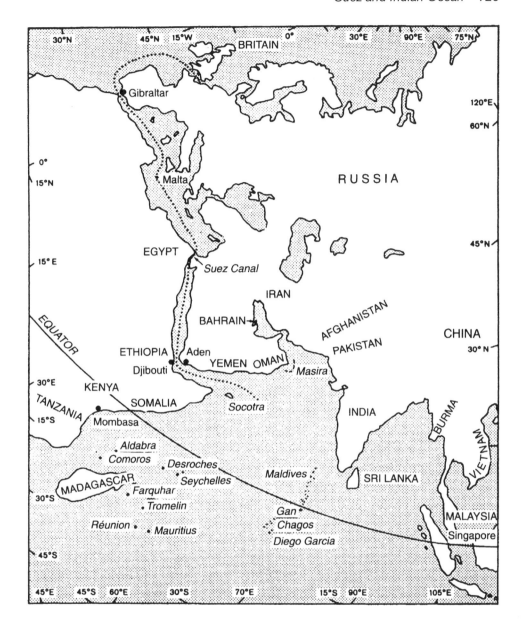

region presented its own problems. Before the airfield was built, 1,400 people of Mauritian origin, who had come to Diego Garcia as plantation workers, were evacuated to Mauritius; in 1982, Britain agreed to pay compensation to these '*Ilois*'; meanwhile Mauritius made a claim to the whole Chagos group.

Réunion remained an overseas *département* of France, which also had base facilities at Djibouti *(35)*. France had granted independence to Madagascar in 1960 and to the Comoros in 1975, but the people of one of the Comoro islands, Mayotte, had insisted on remaining under French protection.

The Indian Ocean's small island states did not always preserve the image of a tranquil tropical paradise. In Seychelles the regime installed by a coup in 1977 survived later attempts at countercoups by calling troops from Tanzania to its aid. In 1988, India sent troops to help the Maldives fight off an attempted coup in which mercenaries from Sri Lanka were employed. In 1989 a French force removed a band of French and Belgian mercenaries who had made themselves masters of the Comoros; in 1995, France had to act again to prevent a repetition of this. In 1997 two of the three main islands, Anjouan and Mohéli, declared independence from the government on the third, Grande Comore; an agreement on greater autonomy for each island reunited the country four years later.

The islands of the Maldives were hard hit by the Asian tsunami of 2004 *(64)*. With a maximum elevation of 8 feet above sea level, they are particularly vulnerable to the effects of global warming; even a small sea-level rise would force evacuation of many islands.

The 1980–8 war between Iraq and Iran, which led to attacks on Gulf oil shipping, and Iraq's invasion of Kuwait in 1990 *(48)* turned the major powers' minds towards co-operation in the region, instead of confrontation. Soviet warships joined American and British in protecting tankers against Iranian attacks; Soviet votes at the United Nations backed the ousting of Iraq's forces from Kuwait by the Americans and their allies. Then the breakup of the USSR meant that the early 1990s saw an Indian Ocean in which there was no longer a Soviet naval presence. But wars and other conflicts were enough to keep US naval and air forces in the region; both Iraq (through 2003) and Afghanistan (2001) were bombed by planes based at Diego Garcia.

43 Israel and Arabs I

In an area formerly part of the Turkish empire, Syria and Lebanon were taken over by France, and Palestine and Transjordan by Britain, after the 1914–18 war – under League of Nations mandates. The Palestine mandate provided for the creation of a modern Jewish 'national home' in the biblical homeland, without prejudice to other communities' rights – a difficult aim. Since the 1890s the Zionist movement had promoted Jewish settlement in Palestine. The persecution of Jews in Nazi Germany increased their migration – though Britain refused to allow free entry of Jewish refugees – and the Arabs began to attack their settlements. The Nazis' massacre of millions of Jews during the 1939–45 war brought Zionism wider support, particularly in America; Holocaust survivors struggled to reach Palestine; the British, unable to stop Arab–Jewish fighting, took the problem to the United Nations.

British Palestine at first included what is today Israel (including much of the Golan Heights), Jordan, the West Bank and the Gaza Strip. In the early 1920s, Britain gave the Golan region to France and designated the region east of the Jordan river as the new territory of Transjordan; what remained retained the name Palestine. In 1947 the UN approved a plan to partition this area (a quarter of the original mandate, then containing about 600,000 Jews, 1,100,000 Muslim Arabs and 150,000 mainly Arab Christians); the Arabs rejected this plan. In 1948 the British left. The Jews proclaimed their state of Israel; it was attacked by all the neighbouring Arab states and Iraq, which undertook to destroy it.

After the Israelis managed to push back the invaders, UN mediation secured armistices in 1949. Egypt kept a hold on Gaza. Transjordan (independent since 1946) annexed the 'West Bank' areas of Judaea and Samaria and half of Jerusalem, renaming itself Jordan. About 700,000 Arabs from the area that was now Israel became refugees in Gaza, Jordan and Lebanon; about 700,000 Jewish refugees from Arab countries came to Israel in its first years. The Arab governments refused to make peace, or recognize Israel, or let its ships use the Suez Canal or the Gulf of Aqaba. On the frontiers, Arab raids and Israeli counterattacks continued.

In 1956, Egypt's expropriation of the Suez Canal Company set off an international crisis. In October, Egypt, Syria and Jordan formed a new alliance against Israel. Israel then attacked and defeated Egypt's army in Sinai. The British and French governments pressed Egypt to let them take control of the canal; their declared aim was only to protect shipping, but they wanted to strike a blow at Egypt, which had been working against them all over the Arab world. The extent of their collusion with Israel became clear later. When Egypt rejected their demands, they destroyed its air force and then captured Port Said. The UN called for British, French and Israeli withdrawals (these were completed by April 1957) and approved the sending of an emergency force (UNEF), which policed the Egypt–Israel

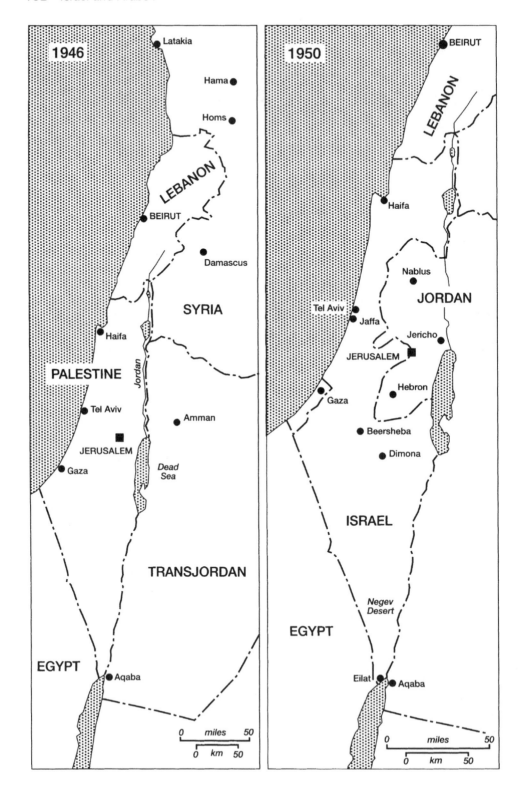

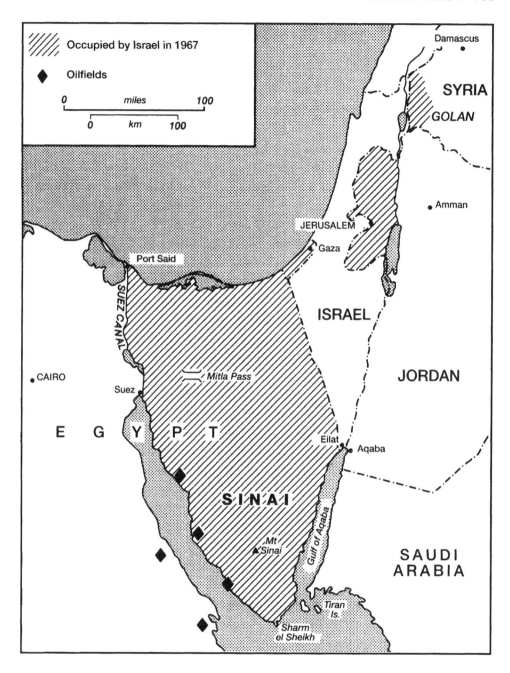

border for ten years. Israeli ships could now use the Gulf of Aqaba; UNEF maintained a post at Sharm el-Sheikh.

In May 1967, Egypt moved its army up to the Israeli border, demanded the removal of UNEF and announced a new blockade of the Gulf of Aqaba. Israel's appeals to the UN brought it no reassurance, and in June it attacked Egypt, Jordan and Syria. In this 'Six-Day

War' Israel captured the West Bank, Gaza and all of Sinai, and the Golan heights, from which Syria had shelled the Galilee lowlands.

The Suez Canal was now a 'front line', closed to all traffic, but the Gulf of Aqaba was again open to Israeli ships. Israel had acquired better defence lines than before 1967 (when even Tel Aviv was within range of shellfire from Jordan); but it had also acquired large Arab populations in its 'occupied territories'. With all of Palestine in Israeli hands, the Palestinian Arabs developed a new resistance. Rival groups combined to form the Palestine Liberation Organization (PLO) – although various Arab states still backed various groups, and terrorist acts in western countries offset the PLO's efforts to win international sympathy.

In 1970 the PLO forces in Jordan challenged its government, were defeated, and withdrew to Lebanon. From then on it was Israel's northernmost areas that were most persistently attacked.

In October 1973, Egypt and Syria attacked Israel while it was observing the annual Yom Kippur rites. The Syrians were soon defeated, and the Israelis advanced to within 25 miles of Damascus. The Egyptians had crossed the Suez Canal, but a counterattacking Israeli force, crossing the canal near Ismailia, turned south and encircled an Egyptian army east of Suez town. This 'Yom Kippur War' aroused fears of a direct American–Soviet confrontation. Arab states cut oil deliveries, setting off an 'oil shock' that affected the whole world economy *(3)*.

The UN obtained ceasefires, and sent a second UNEF to Sinai and another force to man an Israeli–Syrian buffer zone. Israeli withdrawals in 1974–5 released the trapped Egyptian army and permitted Egypt to occupy both banks of the canal (which was reopened in 1975 after eight years' closure), with UNEF-2 manning a buffer zone in Sinai. Egypt, disenchanted by the ineffectiveness of the Soviet arms aid it had received, became increasingly ready to talk about peace. In 1977, President Anwar Sadat made a dramatic visit to Israel and started a series of talks.

44 Israel and Arabs II

In 1978 the Egyptian and Israeli leaders met as President Carter's guests at Camp David, the rural presidential retreat north-west of Washington. In 1979 they signed a peace treaty based on the 'Camp David agreements'. The benefits it brought Egypt included large amounts of American aid as well as the recovery of Sinai. Israel, for the first time in its existence, was at peace with one of its immediate neighbours.

Resenting the peace treaty, other Arab governments broke off relations with Egypt (but most of them had restored relations by 1987). The Arab League suspended Egypt's membership (but readmitted it in 1989). The Soviet Union (which, after hailing Israel's birth in 1948, had goaded on its enemies in 1967 and 1973) also voiced its anger. Egypt withstood the pressure; Sadat was murdered by extremists in 1981, but the treaty held. Israel handed back the greater part of Sinai in 1980, and completed its withdrawal by 1982. A Soviet veto forced the removal of UNEF-2, but an American-organized substitute, the 'multinational force and observers' (MFO), was installed in 1982 along the Egypt–Israel border and the Gulf of Aqaba.

The Israel–Egypt treaty provided for negotiations about giving the 'occupied territories' autonomy, but these became deadlocked. Meanwhile Israeli settlements in the West Bank multiplied. In 1975, 3,000 Jews lived there; by 1985, 45,000; by 1995, 145,000 (and east Jerusalem, solidly Arab in 1967 because Jordan had excluded Jews from the territory under its control, became half Jewish). Some settlers were devoted to the idea that Israel should embrace all of Palestine. In 1987 the Arabs' frustration broke out in an *intifada* ('uprising'), a campaign of sniping, stone-throwing and arson which led to new repressive measures.

The PLO leadership had been recognized by many Arab and other governments as, in effect, a government-in-exile. For years, its statements about wishing to live in peace with Israel were ambiguous, yet went too far to suit the terrorist groups encouraged, in some cases virtually owned, by Iran, Iraq, Libya or Syria. In 1990, when the PLO leaders praised Iraq's aggression against Kuwait, they thereby broke with both Egypt and Syria, and so angered Saudi Arabia and the Gulf states that 600,000 Palestinians who were living there were forced to leave.

In late 1991 the Americans (their hand strengthened both by their expulsion of Iraq from Kuwait and by the eclipse of the Soviet Union) got Israeli–Arab bilateral contacts going. Despite many setbacks, they kept up the pressure. Jordan – which had already renounced its claims to the West Bank – was induced to end its state of war with Israel, and then to sign a peace treaty, in 1994. Israel and Syria were deadlocked; Syria wanted to have the whole Golan region handed back in one piece, and rejected any conditions, compromises or phasing; and while this deadlock lasted, there could be no agreement between Israel and

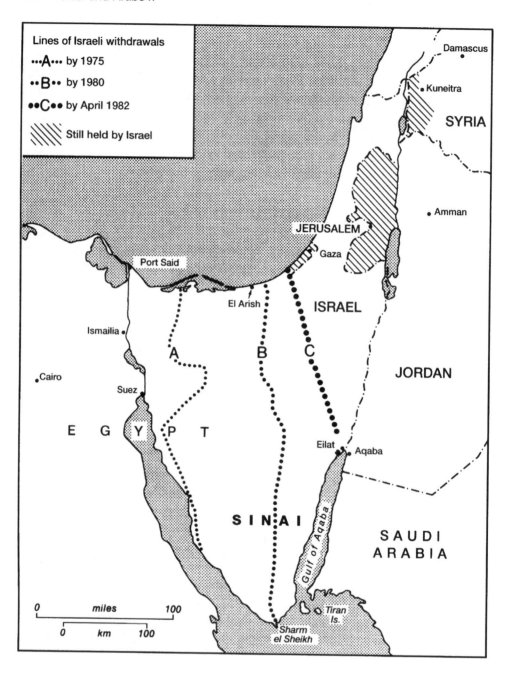

Lines of Israeli withdrawals

···**A**··· by 1975

··**B**·· by 1980

··**C**·· by April 1982

///// Still held by Israel

Lebanon, as any Lebanese move needed Syria's approval *(45)*. Negotiation with Syria ended in 2000.

Israeli–Palestinian deals were thrashed out in 1993 at Oslo and in 1995 at Taba, an Egyptian Red Sea resort; each was later solemnized in Washington. The PLO recognized Israel; Israel recognized the PLO as representative of the Palestinian Arabs. As a first move towards Palestinian self-government, control of Gaza and Jericho was transferred in

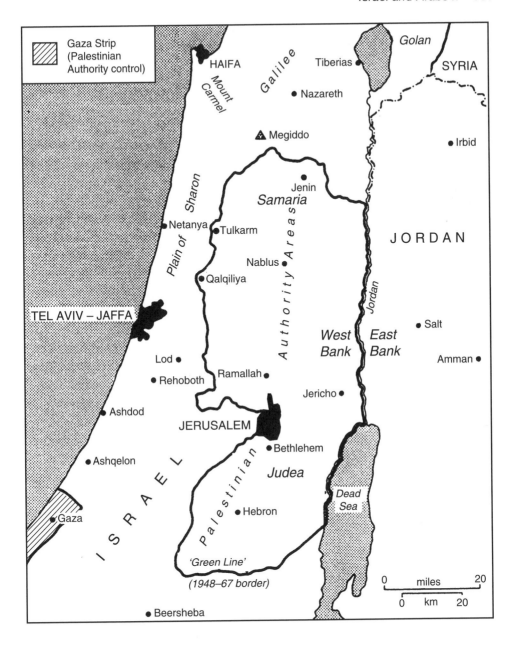

1994. By 1996 the Israelis had handed over other West Bank cities – Bethlehem, Hebron, Jenin, Nablus, Qalqiliya, Ramallah and Tulkarm – and most of the small towns and villages (Hebron, where an Israeli settlement lay in the heart of the Arab city, presented special problems); and a Palestinian Council had been elected, with a *rais* (chief or president) – the PLO leader, Yasser Arafat. The Oslo and Taba agreements envisioned the negotiation of a 'permanent status' pact by 1999; this was not achieved, but subsequent talks at the Wye River Plantation east of Washington and Sharm el-Sheikh (another Egyptian resort) led to the PA controlling or partly controlling about 40% of the West Bank by 2000.

Both sides' extremists tried to disrupt these deals. Denouncing Arafat as a traitor, Hamas, an Islamist group, stepped up terrorist action. In 1995 a Jewish fanatic murdered Israel's prime minister, Yitzhak Rabin; the 1996 election gave Israel a more hardline government.

The Palestinian Authority (PA) government, under the control of Arafat, did not prevent terrorism against Israel. Attacks increased with the 'second *intifada*' that began in 2000 after Israeli foreign minister Ariel Sharon visited the Temple Mount in Jerusalem (location of the ancient Jewish Temple and currently home to important mosques). In 2002, Israel intercepted a ship carrying arms to the Palestinian Authority in violation of the Taba agreement. From 2000 to 2005, hundreds of bombings, shootings and rocket attacks were carried out by members of Hamas, Islamic Jihad, Fatah (part of the PLO) and other terrorist groups, of which the more than 100 'suicide bomber' attacks were particularly deadly.

In retaliation, Israel assassinated numerous leaders of terrorist organizations, but neither military action nor intermittent talks and international pressure succeeded in diminishing terrorist activity. In 2003, Israel began building a 'security fence', a barrier running approximately along the 'Green Line' but incorporating several small areas of dense Israeli settlement in the West Bank (there was already a fence around Gaza). This proved highly effective at preventing the entry of terrorists – the number of incidents declined sharply after a peak in 2002 – if inconvenient for travel and the Palestinian economy.

Starting in 2002, the 'Quartet' of the US, the EU, Russia and the UN sought to restart negotiations with a 'roadmap to peace' that proposed the creation of an independent Palestinian state by 2005, conditional on an end to Palestinian terrorism against Israelis. This was not forthcoming. Yasser Arafat died in 2004, and was replaced by another member of his Fatah organization; but a parliamentary election in early 2006 was won by Hamas. This was seen as a reaction to the corruption of the Palestinian Authority under Fatah; Arafat himself was revealed to have transferred hundreds of millions of dollars to secret bank accounts. The Quartet called on Hamas to renounce terrorism, recognize Israel and accept the peace process, but Hamas refused, and found itself isolated internationally. The US, Canada and the EU cut the flow of aid, bringing the PA government close to collapse.

In 2000, Israel began a series of steps towards 'disengagement' from some of the areas it had held since the 1967 war and the Lebanon invasion by pulling out of its southern Lebanon buffer zone. Its unilateral efforts to withdraw to secure borders and reduce vulnerability to terrorism increased – unexpectedly – after Ariel Sharon (a right-wing former army general) was elected prime minister in 2001. The security fence was planned to separate Israel and the West Bank; and, in 2005, Israel pulled its soldiers and all 8,000 settlers out of Gaza, turning the territory over to the PA. Several small West Bank settlements were evacuated at the same time, as a first step towards a larger disengagement. However, continued rocket fire and other terrorist activity from Hamas (based in Gaza) and Hezbollah (based in Lebanon [45]) led to Israeli attacks on both areas in 2006 and made the new Israeli government reluctant to transfer more territory without better security guarantees.

The years after Oslo saw a greater willingness by Arab states to accept – if not officially recognize – Israel's right to exist. Besides its formal diplomatic relations with Egypt and Jordan, formal ties were arranged with Mauritania in 1999 and trading relationships were established with Oman, Qatar, Tunisia and Morocco in 1994–6 (all but Qatar suspended these ties after the beginning of the 'second *intifada*' in 2000).

In 2006 there were 1.4 million Arabs in Israel, about a million in Gaza and at least 1.5 million in the West Bank. (Thus, in the whole area that had been Palestine in 1947, there were three times as many Arabs as there had been in 1947.) In Jordan there were more than 2 million Arabs of Palestinian origin, so that only by renouncing its claim to the West Bank could Jordan avoid a large Palestinian majority; in other Arab countries there were a million or more, and in other parts of the world half a million.

There were about 5.4 million Jews in Israel, including a million who had come from the former Soviet territories since 1989 and 200,000 in settlements in the West Bank. Of the world total of about 14 million Jews, 6 million were in the United States and more than a million in the EU, and there were large communities in the former Soviet Union, Latin America, Canada, Australia and South Africa.

45 Lebanon and Syria

Being half Muslim, half Christian, Lebanon was given a special status under Turkish rule; under the French mandate *(43)* it was separated from Syria. Its constitutional structure gave the Christians advantages which the Muslims resented; among the Muslims, the Shias resented the Sunnis' privileged position *(28)*; the Christians feared eventual submergence in the surrounding Muslim world; the Druses (a community that broke away from Islam a thousand years ago) distrusted all other Lebanese. After 1970 the Palestine Liberation Organization (PLO) based its forces in south Lebanon *(43)*, and its attacks on Israel brought reprisals. Under all these strains, the Arab world's most liberal society collapsed in bloodshed.

In the confused civil war that began in 1975 the main conflicts were between the Palestinians, with some leftist, mainly Shia, allies, and the biggest Christian sect, the Maronites. A sort of partition took shape, the Maronites holding a zone to the north of the capital, Beirut, while the PLO and its allies controlled the south and the far north. In 1976, Syria sent in troops and halted the fighting; at first many Lebanese were relieved, but suspicions grew when the Syrian troops settled in.

In 1978 the Israelis, seeking to stop PLO raiding and missile-firing across the frontier, occupied the area south of the Litani river. They withdrew when a UN force arrived, but they handed over a 'security zone' along the border to a local militia (the South Lebanon Army or SLA) which had been holding a Christian enclave around Marjayoun. Both this militia and the PLO harassed the UN force, which could not even create a continuous buffer zone.

In 1981 the Syrians attacked East Beirut and other Christian-held towns. In 1982 the Israelis returned, driving the PLO out of the south and trapping 11,000 of its men in West Beirut. After a two-month siege the PLO men were evacuated. A French–Italian–American peacekeeping force, with a small British contingent, was sent to Beirut. Israelis and Syrians faced each other along a line running north-west from Mount Hermon.

In 1983, Syria forced the PLO out of north Lebanon; the Israelis pulled back to the Awali river; fighting between factions began again.

'Suicide bombers' killed 300 US and French soldiers, and in 1984 the peacekeeping force left Beirut. In 1985 the Israelis completed their withdrawal; Shia militiamen took over most of south Lebanon. In Beirut, Europeans and Americans were murdered or taken hostage, mostly by Iran-backed Shia gangs (some hostages were held until 1992).

Lebanon's small army had long been brushed aside, but in 1989 its Christian commander, General Aoun, tried to oust the Syrians. Beirut was shelled by both sides; most of its inhabitants fled. In 1990 the Syrians finally defeated Aoun, ending the civil war. They then tightened their grip on Lebanon.

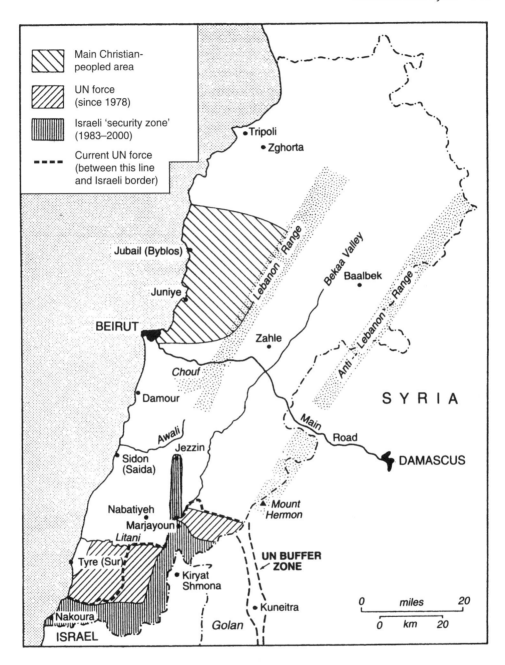

In the south two Shia militias – Amal, backed by Syria, and Hezbollah, backed by Iran
– were fighting each other, and PLO bases had reappeared near Sidon; the Druses kept
their stronghold in the Chouf hills (from which they had driven out 175,000 Christians);
Maronites still held part of the former Christian zone north of Beirut; the Israeli-backed
militia held the 'security zone' along the southern border; but the 40,000 Syrian soldiers in
Lebanon controlled the rest of the country.

In 1991, Lebanon signed a 'treaty of brotherhood' with Syria. Elections were held the following year, based on the 1989 'Taif accord', a constitutional reform that ended the Christians' political predominance. In protest, 80% of Christians boycotted the voting, but many soon accepted the new situation and others emigrated; they now make up only about 40% of the population. A docile pro-Syrian government, in which some Christians and Druses participated, was formed. Beirut began to revive; most of the country enjoyed a Syrian-guarded peace.

Hezbollah, the Shia movement backed and paid by Iran, took part in the Lebanese elections but did not join in the new Syrian-steered consensus. Its guerrillas' continuing attacks on Israel, and on Israelis in the 'security zone', brought large-scale Israeli retaliations in 1993 and 1996. It was reasonably clear that the Syrians were capable of restraining Hezbollah but unwilling to do so. They flatly rejected an Israeli offer in late 1996 to abandon the 'security zone' in return for the disbanding of the Hezbollah guerrillas in southern Lebanon. For the Syrian government, it was important that there should be no peace between Lebanon and Israel until Israel met Syria's own conditions for peace – primarily, the return of the Golan heights *(44)*.

Holding the security zone brought casualties to both the Israelis and the SLA without preventing Hezbollah's short-range rocket strikes on northern Israel. Israel decided to withdraw in 2000, and many members of the SLA escaped across the border.

In 1992, Rafik Hariri, a billionaire with a successful business record, became prime minister. He was able to obtain international aid and use his own money towards reconstruction and reform. Though he built up Lebanon's debt, he also cultivated ties with Europe and the US and undermined Syria's influence. Pressured by Syria and its allies (including Lebanon's president), he resigned in 2004 to support the opposition, but was assassinated the following year. Popular anger at Syria then forced the Syrian army to leave Lebanon (the 'cedar revolution').

Syria's withdrawal left Hezbollah as the strongest power in the country. Alone among the rival Lebanese militias, it had retained its arms at the end of the civil war, and it continued to receive financial and military support from Iran and Syria. The Lebanese army was too weak to restrain it, and the UN peacekeeping force in the south was ineffective; Hezbollah continued its rocket attacks on Israel. In July 2006 it kidnapped two Israeli soldiers along the border. This sparked a large-scale Israeli effort to defeat Hezbollah and force it out of the border region. Though they disabled a substantial part of Hezbollah's arsenal, the Israelis were unable to accomplish either of their aims. Hezbollah's use of 'human shields' for its forces (intentional location of rocket launchers and other military targets within civilian population clusters) and longer-range missiles that could strike Israeli cities increased the number of deaths. An uneasy ceasefire was established after a month of fighting; a more robust UN force was promised.

46 Arabia

For centuries, the Arabian peninsula and the Arab lands north of it were dominated by Turkey. The Turkish empire collapsed after the 1914–18 war. Parts of the region came under British or French control, but Saudi Arabia and Yemen emerged as independent states.

Saudi Arabia had taken its present shape by 1934. From a base among the Wahabi (puritan) Arabs of the Nejd, the Saud family had extended its power over the Hasa, Hejaz and Asir regions. From 1939 on, the oilfields in Hasa turned a desert country into the world's biggest oil exporter *(3, 41)*. The Saudis often subsidized other Arab states (and, for a long time, the PLO *[43]*). After the big oil price rises in the 1970s, their decisions about output, and about investment of their oil revenues, affected the world economy. Saudi Arabia remained a traditionalist country – the guardian of Islam's holiest place, Mecca *(28)*. Pilgrimages to Mecca were disrupted in the 1980s by Iranians who staged demonstrations there; after violent clashes, the Saudis limited the number of pilgrims from Iran, which launched bitter anti-Saudi propaganda campaigns.

At the time of the 'second Gulf war' *(48)* the US based forces in Saudi Arabia. Ten years later, most of those responsible for the September 2001 terrorist attacks in the US were Saudis, affiliated with al-Qaeda and its Saudi leader Osama bin Laden, who condemned the presence of 'infidels' in his birthplace. There were also numerous bombings of oil-processing facilities and American and British targets in Saudi Arabia and Yemen. Both countries were quick to declare support for the US-led 'war on terror', but Saudi Arabia refused to permit the use of its territory for the invasion of Iraq in 2003 *(48)*. The US then withdrew most of its forces and expanded its base in neighbouring Qatar.

Yemen, a relatively fertile region, endured a long civil war from 1962 to 1969. The Saudis backed one side; Egypt sent 50,000 soldiers to fight for the other, and the USSR sent it arms; the war ended inconclusively, and several coups and revolts followed. Discoveries of oil in the 1980s improved the country's prospects. In South Yemen, British rule ended in 1967. Britain had held the port of Aden since 1839 *(42)* and had later created a protectorate in the area. When the British withdrew, a far-left group won the struggle for power and Aden became, for a time, a Soviet naval base. There were frequent quarrels among the new rulers in Aden, notably in 1986, when fierce fighting there caused the hasty evacuation of 4,000 Russians to Djibouti. In the late 1980s, Soviet subsidies for South Yemen stopped, leaving it virtually bankrupt. In 1990 the two Yemeni states agreed to merge. They failed, however, to merge their armies; when the south tried to break away again in 1994 there was a brief war, ending with the northern forces' capture of Aden. Meanwhile, a million Yemenis working in Saudi Arabia had been expelled because Yemen had praised Iraq's occupation of Kuwait *(48)*.

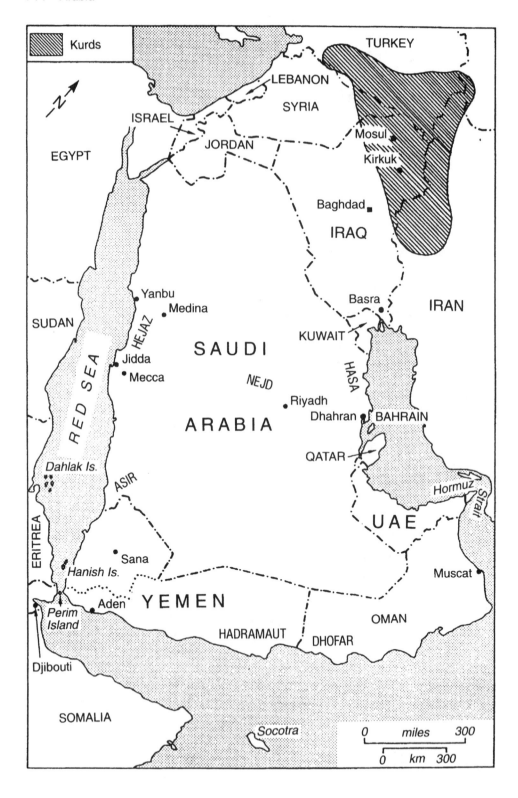

Oman (formerly Muscat and Oman), after a period of British protection, asserted its full independence in 1970. During the 1980s there were revolts in its Dhofar area, which South Yemen fostered and Britain helped to suppress. In 1981, Oman, Saudi Arabia and the four small Gulf states – Kuwait, Bahrain, Qatar and the UAE *(47)* – formed the Gulf Co-operation Council.

In the north, after 1918, France controlled *Syria* and *Lebanon*, and Britain controlled *Palestine* and *Transjordan*, until the 1940s *(43–5)*. The British withdrawal from Palestine in 1948 was followed by the creation of Israel and the annexation of part of Palestine by Transjordan, which was renamed Jordan. The British had given *Iraq* independence in 1932 (they returned in 1941 to oust a pro-Nazi group which had seized power, but soon left again). In 1955, Iraq joined with Britain, Iran, Pakistan and Turkey in the Baghdad Pact, an American-backed alliance intended as a barrier between the USSR and the Middle East. In 1958 a military coup ousted Iraq's pro-western government, and Syria was merged with Egypt in the United Arab Republic (UAR). These events alarmed Lebanon and Jordan, and United Nations observers and American and British forces were sent there for a few months.

Syria left the UAR in 1961 (Egypt went on using the name until 1971). Iraq withdrew from the Baghdad Pact; the other members changed it into the Central Treaty Organization (CENTO), which expired in 1979. A bitter enmity developed between Syria and Iraq, although both regimes claimed to be socialist and obtained Soviet arms and aid. The bitterness was increased when Syria backed Iran against Iraq in their 1980–8 war. When Iraq occupied Kuwait, Syria even sent troops to join the US-led line-up against it.

47 Gulf States and Iran

Now that Persia is called Iran, what used to be known as the Persian Gulf is often just called 'the Gulf'. (Confusingly, 'Gulf', in oil-industry jargon, sometimes refers to the Gulf of Mexico.)

Saudi Arabia has been called a Gulf state, although its Red Sea coast is longer; so has Oman, which is outside the Gulf, except for a small enclave at the Hormuz Strait. But, strictly, the 'Gulf states' are four small ones: *Kuwait, Bahrain, Qatar* and the *United Arab Emirates* (UAE). All were formerly under British protection by treaty, as was Oman.

Kuwait became fully independent in 1961; the other three in 1971. The UAE had been formed by Abu Dhabi, Dubai, Sharjah, Ajman, Umm al Qaiwain, Ras al Khaimah and Fujairah. These seven sheikhdoms were formerly known as the Trucial States, or Trucial Oman; in 1853 they had signed a 'Perpetual Maritime Truce' with Britain.

In 1952, Britain was involved, as the protecting power of Oman and Abu Dhabi, when Saudi Arabia claimed and occupied the Buraimi oasis area on the border between those two states; after long wrangling, the Saudis withdrew in 1955. Between Kuwait and Saudi Arabia, until 1966, there was a 'neutral zone' which they jointly controlled; they then divided the area, but continued to share the oil revenues from it. An old territorial dispute between Bahrain and Qatar over the Hawar islets was settled in 2001, but parts of the borders between Saudi Arabia and its neighbours remained uncertain.

Elected or partly elected parliaments with quite limited powers have been allowed or approved in most Gulf states, but nowhere has the monarch ceded real authority. In some cases women have been allowed to vote.

The four Gulf states joined Saudi Arabia and Oman in forming the Gulf Co-operation Council in 1981. All six GCC states had been enriched by the oil price rises of the 1970s *(3, 41)*. With the oil boom, immigrants flooded in; these included many Palestinian Arabs, but the majority came from southern Asia. On the Gulf's Arab side there are small Iranian communities, and larger ones of Shia Arabs; Iran's influence on them has, at times, troubled the GCC states' Sunni rulers *(28)*.

Iran, which had long maintained a claim to Bahrain, renounced it in 1970. In 1971, Iran occupied Abu Musa and the two Tumbs islands just west of Hormuz, which UAE member states claimed. Later it was agreed that Iran could man a base there, while the UAE retained administrative control; the question of sovereignty was left unresolved. However, in 1992, Iran claimed sovereignty and ejected UAE officials from these islands.

In Iran's population of 70 million, not much more than half are Farsi-speaking Persians. At least a quarter are Azerbaijani (or Azeri) Turks. Near its western frontiers there are Kurdish and Arab minorities, near its eastern ones Turkmens and Baluchis.

In 1941, British and Soviet forces, ousting a government that favoured Nazi Germany,

occupied Iran (which had stopped calling itself Persia in 1935 – besides being the country's indigenous name, the word 'Iran' comes from the same root as 'Aryan'). American and British war supplies were sent to the USSR across it. When the 1939–45 war ended, the USSR continued to occupy north-western Iran, setting up puppet Azerbaijani and Kurdish regimes there. The Soviet forces left in 1946, and Iranian rule was restored; but Iranian fear of the USSR was heightened. When Iran asserted itself against western economic power in 1951 by seizing the British oil refinery at Abadan, the US and Britain arranged a coup to overthrow the prime minister. In 1954, British and US oil companies resumed production, and Iran joined the western-backed Baghdad Pact in 1955. Later, it was helped to build up its armed forces by the Americans, who hoped that Iran would shield the Middle East from Soviet pressures.

In the 1970s there was growing opposition to the regime headed by the Shah. His enthusiasm for education and social reform angered the Shia Muslim mullahs; his pro-American policy angered leftists; joining hands, they fomented mob violence. By the end of 1978, Iran was in turmoil; its oil exports were halted, and this set off the second big wave of worldwide oil price rises *(3)*. In 1979 the Shah was forced into exile. Power passed to the

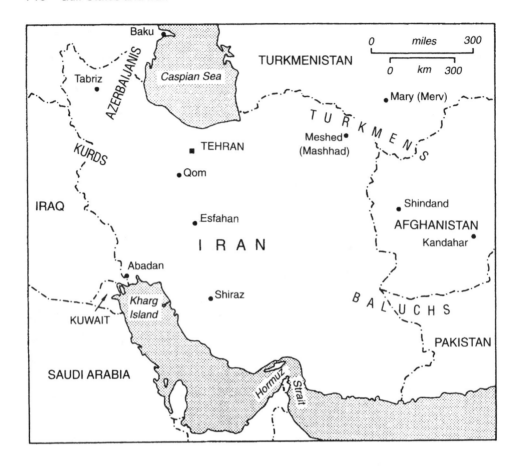

mullahs; they crushed their former leftist allies, and imposed a regime which became notorious for religious persecution.

They denounced the Soviet invasion of Afghanistan *(50)*, which sent 3 million Afghans fleeing into Iran. But they were more interested in hounding the Americans. The US embassy in Tehran was seized, and its staff were held hostage for more than a year. This violation of a basic rule of international relations caused widespread outrage. Iran's consequent isolation encouraged Iraq to invade it in 1980 *(48)*.

During the war that ensued, Iran failed to turn the Shias of southern Iraq against their rulers; elsewhere, however, its appeal to Shia loyalty sometimes prevailed over old Arab suspicions of Iran. It armed, paid and directed some of the Shia groups in Lebanon which, as well as helping to wreck the country by sectarian fighting, specialized in seizing Europeans and Americans as hostages *(45)*.

Iran made ruthless use of assassins in Europe and elsewhere. In 1989 it told all Muslims that it was their duty to murder an author living in Britain who, according to the mullahs, deserved death as a 'blasphemer'; this *fatwa* is still in force *(28)*. Yet, in the 1980s, while Iran's ruling mullahs were enforcing a strict 'Islamic' orthodoxy and thundering against Jews and against the United States, they secretly bought Israeli arms through the US (on which they could put pressure because Iranian-backed groups in Lebanon were holding Americans hostage); this emerged during the 'Irangate' hearings in Washington in 1987.

One effect of its new rulers' actions was that Iran got very little international sympathy when Iraq invaded it. As that war dragged on, Iran seemed insistent on continuing the bloodshed, although the Iraqis, having been driven back all along the frontier, were asking for peace. And it was Iranian attacks on neutral shipping in the Gulf that brought American, Soviet, British and other warships there to protect tankers.

Neither the ending of the Iran–Iraq war nor the concentrating of international anger on the Iraqis when they invaded Kuwait sufficed to remove all fears about Iran's intentions. In 1995 the US stopped all its direct trade with Iran. However, as Iran (unlike Iraq) was freely exporting its oil, it could offer lucrative deals to Europeans and others. Disputes about whether it was wise to do business with Iran caused friction between the Americans and the Europeans and the Japanese.

Power in Iran is shared between unelected conservative Islamic authorities headed by the Supreme Leader (a Shia ayatollah) and an elected president and parliament. A comparatively liberal president elected in 1997 sought reforms, including greater press freedom and some warming of relations with the US, but his efforts were soon blocked by the conservatives.

Iran's pursuit of nuclear weapons was of growing concern. In 2002, Russia began building Iran's first reactor (ostensibly for electric power, despite the country's oil wealth); in 2003, the International Atomic Energy Agency demanded inspections under the Non-Proliferation Treaty (NPT *[4]*). Iran at first co-operated, but after the 2005 election of a hardline president who called for the destruction of Israel, it resumed enriching uranium and resisted all diplomatic pressure to comply with the NPT. By 2006 even Russia and China had begun to participate in anti-enrichment efforts by the US and the EU.

48 Iraq's Wars

The 1980–8 war between Iraq and Iran followed a series of disputes between them, particularly over the Shatt al-Arab, the waterway through which the Euphrates and the Tigris flow into the Gulf. Iraq had also tried to stir up unrest among the Arabs in Iran's Khuzestan province; similarly, Iran wooed the Shias of Iraq (28); in this, neither had much success.

Iraq's ruler, Saddam Hussein Takriti, invaded Iran in the mistaken belief that its army was too shaken by the Shah's fall (47) to offer much resistance. By 1982 the invaders had been pushed back to the frontier in most sectors. Saddam sued for peace, but the Iranian rulers said they would fight on until he was brought down. They had blocked Basra, Iraq's only port; Iraq could export oil by pipeline, but all its imports had to come in by costly road transport. Although Saddam obtained subsidies from Kuwait and Saudi Arabia, Iraq's economy was hard hit.

Iran's ruling mullahs proclaimed a holy war (*jihad*) and used 'human sea' tactics, sending forward wave after wave of men, and often children, to be mown down. (Iraq's war dead exceeded 250,000, but Iran's were three times as many.) By 1986 the Iranian forces had crossed the Shatt near its mouth, capturing Fao.

This 'Gulf war' spread to the Gulf itself in 1984, when Iraq began air attacks on Kharg island, Iran's main oil terminal. Iran then attacked ships using Kuwaiti ports, refusing to regard them as neutral. The 'tanker war' escalated; as Kuwait had no export pipelines, its oil exports were particularly imperilled. In 1987 the United States took over some Kuwaiti tankers and provided escorts for them. By then British, French and other mine-sweepers were in the Gulf; the 75 warships there included Soviet ones, protecting Soviet tankers chartered by Kuwait. Iran had threatened to close the Gulf to all shipping – using missiles deployed near the Hormuz strait – but the maritime powers' angry reactions deterred it.

In 1988, United Nations pressure on Iran secured a ceasefire, to Saddam's relief. In 1990, after making a second disastrous mistake by underestimating the reactions to his swoop on Kuwait, he had to appease Iran. He pulled his troops out of the small areas in Iran which they still held, and agreed to share control of the Shatt, abandoning his primary war aim of obtaining full control of the waterway. Thus, his eight-year war had left an impoverished Iraq with no gains of any kind.

Iraq was also left with a large army and a frustrated, dictatorial and still ambitious ruler. Its weaker Arab neighbours eyed it anxiously. It became known that Iraq was producing poison gas at Samarra (it had used gas against Iran and its own Kurds [49]), and it was detected trying to buy components for long-range 'super-guns' and for nuclear weapons. Soon Saddam was pressing Kuwait to hand over some border areas and oilfields.

In 1961, Iraq – then ruled by General Qasim, who had seized power in 1958 – had asserted a claim to the whole of Kuwait. The other Arab states opposed this claim (Kuwait had never been part of an independent Iraq). The British force sent to protect Kuwait in 1961 was soon replaced by an Arab League force, which remained there until Iraq renounced its claim in 1963.

In August 1990, Saddam's army attacked and overran Kuwait, and he announced its annexation. The UN Security Council ordered Iraq to withdraw, imposed a trade embargo and other sanctions, and authorized the use of 'all necessary means' to make Iraq withdraw if it had not done so by mid-January 1991. The Arab League approved the sending of troops to protect Saudi Arabia, in response to urgent Saudi appeals.

Under the trade embargo, the pipelines carrying Iraqi oil across Turkish and Saudi territory were shut, and the Saudis clearly risked a retaliatory attack. An American-led coalition mustered air, ground and naval forces in Saudi Arabia and the Gulf. They came from America, Britain, Canada, France, the USSR and 20 other countries; among these, notably, were seven Arab states (including Egypt and Syria) and four other Muslim states (including Bangladesh and Pakistan).

When the January 1991 deadline arrived, air attacks on Iraq began. In February, Kuwait was liberated in four days by American, British, French, Egyptian, Syrian, Saudi and other Arab forces. The fleeing Iraqi troops carried off all the loot they could, and set many of Kuwait's oilfields on fire; it took nine months to extinguish these fires.

During this brief 'second Gulf war', Saddam fired Scud missiles into Israel as well as into Saudi Arabia. Israel did not hit back; it thus defeated his plan to drag it into the war so that he could pose as the champion of Arab grievances against Israel, diverting attention from his aggression against an Arab state. Among the Arabs, only Jordan, Sudan, Yemen and the PLO *(43)* had hailed his swoop on Kuwait; and all of these except Sudan switched to condemning it when they felt the effects of other Arab states' anger.

Defeated, Iraq had to admit a UN monitoring commission whose task was to discover and destroy its stocks of, and capacity for making, nuclear, biological or chemical 'weapons of mass destruction' (WMDs). The ban on Iraqi oil exports, imposed in 1990, was to be maintained until the commission completed its work. However, Iraq was offered the opportunity to sell a limited amount of oil, provided that part of the proceeds went to victims of its aggression against Kuwait and that the rest went to buy food, to be distributed under UN supervision. After protesting for several years that these conditions were unacceptably humiliating, Iraq accepted them in 1996.

Other restraints on Iraq after the 1991 conflict included the imposition of 'no-fly zones' on its air force. The northern zone was created to prevent air attacks on the Kurdish area near the Turkish frontier *(49)*, and the southern covered the Shia region south of Baghdad.

The Iraqis repeatedly tried to obstruct the UN commission, but its inspection teams worked on patiently through 1998, when Iraq refused to co-operate further. Bombing of military targets by the US and Britain through 2003 did not bring Iraq into full compliance with UN requirements, though some progress was made in 2002. The US, Britain and Spain then sought UN Security Council agreement for military action but were opposed by most other members, including Russia, France and Germany, which favoured a greater inspection effort.

Without UN approval but aided by a 'coalition of the willing', the US invaded in March 2003, quickly routing the Iraqi army and seizing Baghdad (Saddam Hussein was not captured until December). Investigation over the next year found no remaining WMDs.

Peace proved elusive. By September 2006 the US still had 130,000 troops in Iraq; Britain and other countries contributed 20,000 more. About forty countries had participated, including predominantly Muslim Azerbaijan and Kazakhstan, but there were no Arab states among them. The coalition had lost 3,000 troops, mostly after the declared end of 'major combat operations' in May 2003, and Iraqi civilian deaths were estimated at more than 40,000 – many from car and suicide bombings. Violence between Sunni and Shia Arabs threatened to escalate towards civil war.

About three out of five Iraqis are Shia Arabs, who dominate the region south of Baghdad; roughly equal numbers of Sunni Arabs and Kurds (also primarily Sunni) are most of the rest. Under Saddam Hussein, the Sunni Arab minority held power. A parliament elected after voters approved a new constitution in 2005 was split largely along ethnic and religious lines, with the main Shia party holding the largest number of seats.

49 Kurds

Nearly all of the more than 25 million Kurds are Sunni Muslims. Their language is of the Iranian group, distantly related to Persian (Farsi). Most of them live in an area that is divided between four countries: Turkey, in which there are at least 14 million Kurds; Iran, with 6 million; Iraq, with at least 4 million; and Syria, with 1 million (there are also some thousands of Kurds in Armenia and Israel). About a third of the Kurds in Turkey have migrated to its western regions, and many have, in varying degrees, become assimilated; a quarter of the members of the Turkish parliament, and a quarter of the 2 million immigrants from Turkey now in Germany, are of Kurdish origin. But the Kurds' heartland is the region stretching from south-east Turkey to northern Iraq and western Iran, whose population is predominantly Kurdish. Rebellious Kurds have repeatedly challenged each of the three governments that share control of this region.

After the 1914–18 war and the collapse of the Ottoman Turkish empire (which had included Iraq and Syria), it was proposed in the 1920 Treaty of Sèvres that, under a League of Nations mandate, Britain should administer an autonomous Kurdistan. Turkey successfully resisted this plan, and the territory that had been proposed for Kurdistan was instead divided between Turkey and Iraq. In later years, several ostensible offers of autonomy were made to Iraq's Kurds by successive governments, but nothing of substance developed.

In the 1920s and 1930s, Turkey suppressed Kurdish revolts. In the 1960s and early 1970s, Iraq was unable to do so until Iran, in 1975, agreed to stop encouraging them. During the 1980–8 war between Iran and Iraq, the Kurdish communities in both countries were denounced for aiding the enemy, and savagely punished. After crushing an uprising in 1988, Iraq's ruler, Saddam Hussein, ordered the destruction of thousands of Kurdish mountain villages, whose inhabitants were forcibly concentrated in new settlements near large towns. He also attacked the town of Halabja with poison gas, massacring Kurdish civilians.

Meanwhile a communist-led movement based in Syria had begun, in 1984, a long guerrilla struggle in south-east Turkey. In the early 1990s a large part of Turkey's army was engaged in campaigning against these PKK (Kurdish Workers' Party) guerrillas. By 1999 it was estimated that 15 years of this conflict had cost 30,000 lives; and the brutal methods used by Turkish soldiers and police when operating in the south-eastern provinces had alienated many of their Kurdish inhabitants, thousands of whom fled from the fighting to Istanbul and other cities in western Turkey. After his capture by the Turkish army in 1999, the imprisoned PKK leader called for a ceasefire.

In 1991, after the Iraqi army's eviction from Kuwait, Iraq's Kurds rebelled again, briefly capturing Kirkuk. They were driven back into the mountains, but the 'coalition' allies who

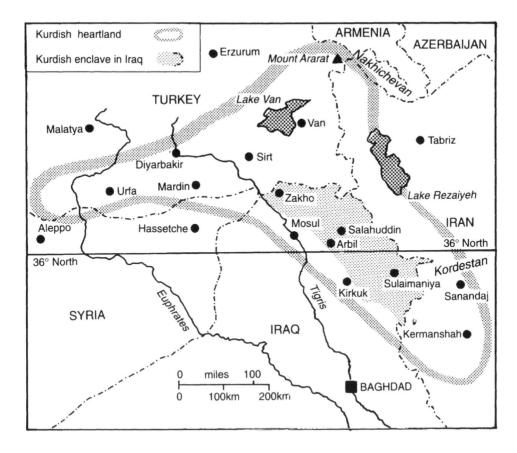

had defeated Iraq sent a small force which stayed there for four months and deterred the Iraqi army from pursuing the Kurds into an enclave designated as a 'safe haven'. Iraq's air force was warned not to fly north of 36° North, and American, British and French aircraft, based at Incirlik in Turkey, policed this 'no-fly zone'; Turkish forces also entered the enclave to attack guerrilla bases.

The Iraqi Kurds failed to create a united administration in their enclave. One faction, the Kurdistan Democratic Party (KDP), controlled its north-west sector; another, the Patriotic Union of Kurdistan (PUK), held the south-east. In 1994–6 they fought each other until American mediation brought a ceasefire.

The Kurdish zone gained greater political autonomy after the overthrow of Saddam Hussein, and the two factions finally agreed on a unified regional government in 2005. The boundaries of Kurdistan were still undefined – the status of oil-rich Kirkuk was a particular problem – but Kurdistan enjoyed greater stability than the rest of Iraq. Kurds were well represented in the new Iraqi government but many still sought independence, a goal particularly opposed by Turkey.

Meanwhile guerrilla activity revived in Turkey after a break of five years. Among the country's Kurds, although support for the PKK was limited, there was resentment of the army's repressive actions and long-standing restrictions on use of the Kurdish language. Starting in 2002, however, Turkey began to liberalize education and broadcasting in Kurdish as part of its effort to gain EU membership *(12)*.

50 Afghanistan

Afghanistan has no linguistic or ethnic unity. Before the 1979–89 Soviet occupation, the main groups in a population of 17 million were 8 million Pathans (or Pushtuns), 4 million Tajiks, 1.5 million Uzbeks and a million Hazaras. The Tajiks, Uzbeks and Turkmens had kindred in what was then Soviet territory *(20)*; the Pathans and Baluchs had kindred in Pakistan. What the frontiers marked out was the (mainly mountainous) area that Pathan rulers based in Kabul had been able to hold against the former encroachments of tsarist Russia, Persia (now Iran), and India's British rulers.

When Pakistan became independent in 1947, the Afghan rulers agitated for its Pathan-peopled North-West Frontier province to be made a separate state, 'Pushtunistan'. This embittered relations with Pakistan; the Afghans' trade with and through it was disrupted; Soviet influence in Afghanistan was thus increased. The Russians built roads and airfields (which were soon to serve their military needs) and exploited the gasfield near Shibarghan, piping the gas into Soviet territory. Two military coups put a fairly pliant pro-Soviet regime in power; but Afghan restiveness grew, and the Soviet officers attached to the Afghan army told Moscow that they could not rely on it.

In 1979 a Soviet army was rushed in; Afghanistan's president was murdered and replaced by a more docile successor. But the Afghans did not accept the Soviet occupation tamely. In the ten-year war that ensued, more than a million people were killed (mainly Afghan civilians); large areas were devastated; over 3 million Afghan refugees fled into Pakistan, another 3 million into Iran.

Many men in the pro-Soviet regime's conscript army went over to the *mujaheddin* (resistance fighters); the Soviet forces had to do much more fighting than they had expected. As Soviet casualties mounted, so did discontent in the USSR itself. Arms and aid reached the resistance from Pakistan and Iran, and also from such varied sources as America, China and Saudi Arabia. By 1988 the guerrillas controlled three-quarters of the country and were even bombarding Kabul. In 1989, under UN supervision, the Soviet forces withdrew.

They had reduced the country to chaos. Some guerrillas began to fight each other; some continued to attack the Marxist regime in Kabul, which, abandoned by the Russians, was hastily abandoning Marxism. By 1992 it had collapsed. The confused struggle that followed was simplified, but intensified, in 1994 by the emergence among the Pathans of a Sunni Muslim fundamentalist movement, the Taliban ('seekers').

This movement originated among refugees in Pakistan, and had some (discreet, and officially denied) Pakistani backing. Its forces took over Kandahar and swept through Afghanistan's Pathan-peopled south. In 1995 they captured Herat and came close to Kabul, taking it the following year. They then faced counterattacks by the Tajik forces based in

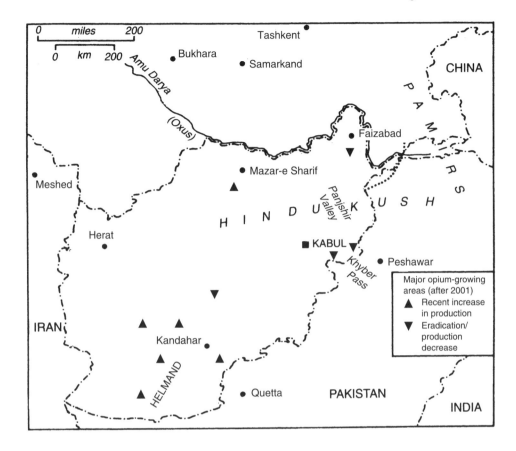

the Panjshir valley and by the forces of General Dostum, which held the Uzbek region and the northern approach to Kabul by the Salang tunnel through the Hindu Kush.

The Tajiks' leader, Ahmad Shah Masoud, had been prominent in the anti-Soviet resistance, whereas Dostum had commanded the Marxist regime's forces. The new alliance of these old antagonists was based on Tajik and Uzbek unwillingness to submit once again to Pathan dominance. The Hazaras shared this feeling; also, being mostly Shia, they feared the Sunni sectarianism of the Taliban – and they were backed by Shia Iran *(28, 47)*. Thus, by 1996, the conflict was directly related to Afghanistan's ethnic, linguistic and religious dividing lines; and all the contending forces had links with, and backing from, neighbouring states.

At first, the Taliban were welcomed in places where they stopped extortion by local gangs and petty warlords; their commanders forbade looting; but their merciless enforcing of a very strict version of 'Islamic law' soon made them less welcome. And the dependence of the puritanical Taliban on funds from opium profits evoked some cynicism. In the 1990s, Afghanistan was producing a third of the world's heroin, from an annual 2,000 tons of raw opium; but in 2000 the Taliban banned it as contrary to Islam, and very little was grown in 2001.

In September 2001 the Taliban refused an American demand to hand over Osama bin Laden, the terrorist leader judged responsible for the attacks on New York and Washington. The US began air strikes in October and forced the Taliban out of Kabul but captured neither bin Laden nor the Taliban leader, Mohammad Omar. Afghan opposition groups formed a temporary government in December supported from 2003 by US troops and a NATO force under European leadership. Elections in 2004 and 2005 brought in a new government that struggled to assert its authority across the whole country, even with continued US and NATO involvement, and opium production surged to twice the pre-2001 level.

51 South Asia I

Britain's Indian empire was the most spectacular feature of the age of colonization in Asia and Africa, and it was the British withdrawal from India that set off the great wave of 'decolonization' *(9, 27)*. Until the 1930s, proposals for India's advance to independence were based on hopes of preserving the unity brought to the whole 'sub-continent' by British rule. But a quarter of undivided India's population, then about 450 million, were Muslims (who had formed the ruling group almost everywhere before the British came). They now insisted on the creation of a separate state, to be called Pakistan; and, to avert a Hindu–Muslim war all over India, a partition was agreed. Even so, before and after the two nations became independent in 1947 there was much bloodshed in Punjab and Bengal, the two provinces where half of the inhabitants were Muslims; these provinces had to be divided. About 8 million Muslims fled from India to Pakistan, and about 8 million Hindus and Sikhs fled from Pakistan to India; half a million people were killed; but 10 million Hindus remained in Pakistan, and 40 million Muslims in India. (There are now 140 million Muslims in India, about 13% of a population that now exceeds a billion.)

Pakistan emerged in 1947 with two 'wings', 1,000 miles apart and each with half of the population. Its Bengalis were soon complaining, with reason, that its western-based government unduly favoured the western wing, which comprised Sind, Baluchistan, the (Pathan-peopled) North-West Frontier Province and the western part of Punjab *(52)*.

Most of the former princely states were quietly absorbed into India. The Muslim ruler of Hyderabad (mainly Hindu-peopled) sought independence, but in 1948 India took over his state. The Hindu ruler of Kashmir (mainly Muslim-peopled) vacillated; a revolt began; Pathans invaded Kashmir from Pakistan; India sent in troops. Muslims backed by Pakistan still held 'Azad Kashmir' in the north-west in 1949, when the United Nations secured a ceasefire and sent observers to maintain it.

Kashmir remained a bone of contention. China sent troops into its almost empty north-east area, Aksai Chin; in 1962, Indian and Chinese forces clashed there. In 1965 new clashes along the 1949 ceasefire 'line of control' led to full-scale fighting between the Indian and Pakistani armies on the Punjab and Sind frontiers; again the UN secured a ceasefire, and it sent more observers to watch over the two armies' withdrawal. From 1989 onwards, Indian forces in Kashmir were trying to suppress Muslim rebels who were demanding independence and contesting the eastern end of the line of control with Pakistan – despite the 20,000-foot elevation and bitterly cold temperatures.

Almost all residents of Pakistani-ruled Kashmir are Muslim, as are two-thirds of those in the Indian-controlled part. An exodus of Hindus from the Srinagar area to Jammu in the south *(53)* left southern Jammu the only Hindu-majority part of Kashmir, while Buddhists are in the majority in much of sparsely populated Ladakh. India favours retaining the

current border as a permanent boundary, but Pakistan opposes a division that leaves most of Kashmir, including Muslim-dominated areas, under Indian control. Kashmiri calls for a plebiscite – including independence as a third option along with union with India or Pakistan – are rejected by both countries.

France and Portugal had held small coastal territories in India. The French ones were handed over to India in 1951 and 1954. Portugal would not agree to a transfer, but in 1961 India took over Goa, Diu and Daman, the last vestiges of European rule in the sub-continent.

Sri Lanka (until 1972, Ceylon), which had been a separate British colony, became independent in 1948. In its population of 20 million, 18% are Hindu Tamils (another 7% are Tamil-speaking Muslims). Although some of the Tamils' ancestors came from India only a century ago, most of the community has been in the island for 2,000 years; but it has natural

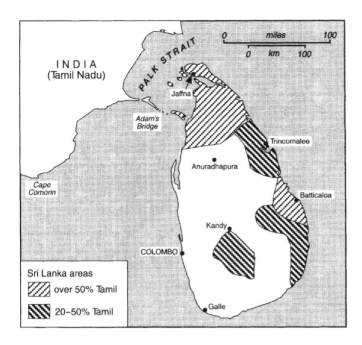

links with the 60 million Tamils in south India. After 1948, laws making Buddhism the state religion and Sinhala the only official language were followed by purging moves which made the civil service, parliament and army almost purely Sinhalese, and populist politicians whipped up pogroms in areas where Tamils were a minority. By the early 1980s many Tamils were backing the guerrilla 'Tigers', who took over the Jaffna area in the north, demanding a separate state, 'Tamil Eelam'.

The guerrillas had received money and arms from south India's Tamils; but in 1987 India, at the request of Sri Lanka's president, sent 50,000 troops to round up the 'Tigers'. They did better than the island's own army had done; but in 1990, also at Sri Lanka's request, they left. There was now little sympathy in India for the 'Tigers'; still less after India's former prime minister, Rajiv Gandhi, was murdered in 1991 by one of their supporters. Fighting continued in the north and the east, with bombings by the Tigers in the capital and elsewhere, until 2002, when talks led to a ceasefire and an agreement on regional autonomy rather than independence in Tamil areas. Subsequent negotiations failed to bring real peace. Two years later, the Asian tsunami caused 35,000 deaths, the majority in areas claimed for 'Eelam', but even this disaster brought no lasting change (as it did in Indonesia *[64]*). An agreement by the government and Tamil and Muslim leaders to distribute foreign aid was not implemented, and fighting intensified once again in 2006.

52 South Asia II

By 1971, Bengali demands for more self-government for East Pakistan had been answered with repressive action, Pakistan's army was fighting Bengali guerrillas, and millions of Bengalis had fled into India. India's army then invaded East Pakistan and defeated the Pakistani forces there. A new republic, *Bangladesh*, was proclaimed in what had been East Pakistan.

Although Pakistan eventually accepted Bangladesh's independence, and they shared a Muslim heritage (83% of Bangladeshis and 97% of Pakistanis are Muslim), their violent parting left many issues in dispute between them. And, although Bangladesh owed its independence to India, they, too, had disputes – particularly over Bengali migration into Assam, and over India's diversions of water from the Ganges. (The huge Ganges delta forms a large part of Bangladesh's territory.)

Meanwhile *India* had laid out new state boundaries, mainly based on linguistic regions: Kerala, for speakers of Malayalam; Tamil Nadu for Tamil; Andhra Pradesh for Telugu; Karnataka (until 1973 Mysore) for Kannada; Gujarat for Gujarati-speakers; Maharashtra for Marathi; Orissa for Oriya; West Bengal for Bengali; Rajasthan, Uttar Pradesh, Madhya Pradesh and Bihar for Hindi and its variants. To accommodate caste differences and indigenous tribal peoples, the last three of these were each split in 2000 to create Uttaranchal, Chhattisgarh and Jharkhand. The part of Punjab left to India by partition was redivided in 1966: the mainly Hindi-speaking part became Haryana, while the Himalayan foothills went to Himachal Pradesh, the residual Punjab being mainly Punjabi-speaking (and 63% Sikh).

In the north-east, Assam was diminished by the creation of states for the peoples of the hill areas: Nagaland, with Kohima as its capital, for the Nagas; Mizoram (formerly the Lushai Hills); Meghalaya, for the Garo, Khasi and Jaintia hill peoples; Arunachal Pradesh (formerly the North-East Frontier Agency). Manipur and Tripura, two former princely states, were given full statehood.

Vast and variegated India, the world's biggest democracy, had to contend with many linguistic, religious and cultural divisions. The worst conflicts involved religious divisions: between Hindus, Muslims and Sikhs; between 'higher' and 'lower' Hindu castes. Terrorist campaigns by Sikh separatists in Punjab led in 1984 to military action against their strong-hold in the Golden Temple of Amritsar (the holiest Sikh shrine), followed by mutinies by some Sikh soldiers, the murder of the prime minister, Indira Gandhi, by her Sikh body-guards, and retaliatory killings of thousands of Sikhs in Delhi. By the early 1990s, Punjab had been quietened by forceful action which proved remarkably popular. But the secular framework which India had maintained since independence was being more severely tested by an upsurge of bigotry among Hindus. One special target was a mosque built in 1528 on

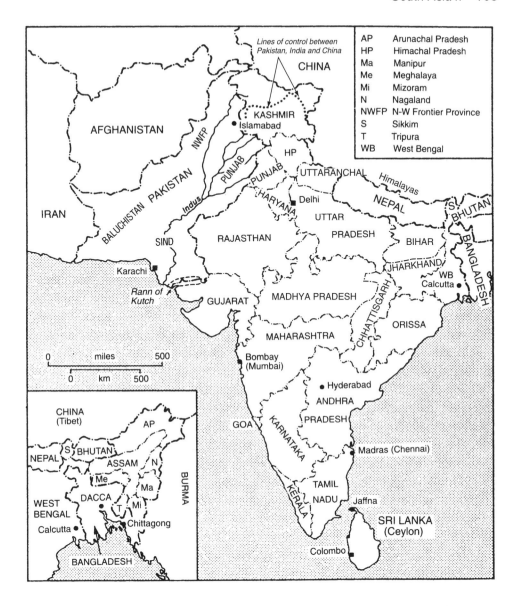

an old Hindu holy site at Ayodhya in Uttar Pradesh; it was eventually destroyed by a mob in 1992, and this set off a new cycle of violence between Hindus and Muslims. The wave of intolerance was the work of brahmins and other high-caste Hindus; some of these also reacted violently to the reserving of more government jobs for lower-caste *sudras* and casteless *dalits* (once known as 'untouchables').

Violence between Hindus and Muslims intensified in 2001 and 2002. India blamed Kashmiri groups for a bombing of the Indian parliament in 2001 and accused Pakistan of supporting terrorism. By 2002, India and Pakistan – both declared nuclear powers since 1998 *(4)* – were close to war, but they managed to agree on a ceasefire in Kashmir in 2003. A few concrete signs of progress followed (including a bus route across the Kashmir border), and talks have continued despite several more terrorist attacks in India.

In *Pakistan* there was recurrent violence between some groups in the Sunni majority and in the Shia minority *(28)*, and, in Karachi, between the local Sindhis and the large *mohajir* (refugee) community which had come from India in or after 1947. The long years of war in Afghanistan added to Pakistan's problems in several ways: too many guns passed into too many hands, and there was much drug trafficking from Afghanistan.

Unlike India, Pakistan has spent many years under military rule. Though a secular state, Pakistan initially supported the Taliban in Afghanistan *(50)* but changed sides after the 2001 terrorist attacks on the US. American forces were allowed to use Pakistani bases to attack Afghanistan, and in return the US ended sanctions enforced after Pakistan's 1998 nuclear tests. In 2004 the country's leading nuclear scientist admitted to having aided the weapons programmes of Iran, North Korea and Libya *(4)*.

In 1985 the prime minister of India, the presidents of Bangladesh, the Maldives, Pakistan and Sri Lanka, and the kings of Bhutan and Nepal met in Dacca (Dhaka) to create the South Asian Association for Regional Co-operation. However, the SAARC grouping remained a very loose one. The smaller members tended to regard India warily, seeing it as heir to the formerly dominant British role in the region.

Nepal could show a long record of success in maintaining its independence. The fighting qualities of the Gurkhas, Nepal's main ethnic element, had impressed the British, and agreements were made permitting them to recruit Gurkha soldiers. In 1947, Nepal agreed to let both Britain and newly independent India recruit Gurkhas. But its relations with India were not always easy. Nepal resented Indian interference in its politics. Later, India came to resent Nepal's links with China. In the late 1980s, when Nepal bought arms from China, its trade negotiations with India became deadlocked and, for a time, India closed the frontier.

Before 1990, Nepal was a monarchy. The first democratic government was replaced in 1994 by a communist one that collapsed the following year, sparking a Maoist rebellion. Subsequent governments were short-lived. In 2001 the crown prince assassinated the king and other members of the royal family, then shot himself. Strikes and violence continued, with the military unable to defeat the Maoists. This led the new king to cancel elections in 2002 and formally take power from the prime minister in 2005. Later that year, the Maoist and other opposition movements agreed to join forces to restore democracy, and in 2006 the king restored the parliament.

In 1947, India had taken over from Britain the handling of the external relations of Sikkim and *Bhutan*; by 1975, Sikkim was one of federal India's states, but Bhutan had asserted its separate identity by joining the United Nations. Later, Bhutan became alarmed about immigration by Nepalese; in the early 1990s it forced some of them out by none too gentle methods.

Both Bhutan and Nepal are among the world's poorest countries in money terms, with per capita GDPs of around $1,000 – considerably lower than that of India (whose technology and manufacturing sectors have created a substantial middle class). But, while Nepal slid into war, Bhutan's king measured success in terms of 'Gross National Happiness' rather than GDP. His government enforced a traditional dress code, kept out television and the internet until 1999, banned tobacco in 2004, and strongly discouraged logging – which might discourage wealthy tourists, the only kind allowed. In 2005, the king announced plans for democratic elections and his abdication in 2008.

53 Himalayas, Tibet, Burma

Tibet, although under formal Chinese suzerainty, was for centuries independent in practice, and ruled by successive Dalai Lamas, who were also the religious leaders of a devoutly Buddhist people. In 1950, however, the communists who had won China's civil war occupied Tibet. Large numbers of Chinese were then settled in Lhasa, the capital, and other parts of the country. Several revolts were suppressed. During one of them, in 1959, the Dalai Lama escaped to India.

Meanwhile China was publishing claims to Himalayan mountain areas long regarded as belonging to India, Nepal and Burma. Its troops entered the Aksai Chin region in north-east Kashmir. There were also clashes on the Indian border north of Assam, where in 1914 the British, who then ruled India and Burma, had fixed (after negotiating with China and Tibet) the frontier called the McMahon Line. The border disputes worsened, and in 1962 the Chinese launched full-scale attacks. In north-east India their forces advanced almost to the edge of the Assam plains. Two months later these forces, whose supply lines were tenuous, withdrew to the McMahon Line; but the Chinese did not renounce their claims in that sector, and in Kashmir they kept their hold on Aksai Chin. When India and Pakistan went to war in 1965, China tried to strengthen Pakistan's hand by threatening to invade India again.

China soon settled its frontier disputes with Nepal and Burma, accepting the Burmese part of the McMahon Line. In 1963 it agreed with Pakistan on a frontier line in north-west Kashmir; Pakistan ceded part of the Karakoram range to China, while India protested that it had no right to do so. In 1982, China and Pakistan opened a road across the range, linking Kashgar with Gilgit and thus with Islamabad. In northern Ladakh, Indian and Pakistani forces repeatedly clashed on the mountains around the Siachen glacier. In 1993, India and China agreed on troop reductions along their border; but they maintained their conflicting claims. New talks aimed at resolving border disputes in north-east India and Kashmir began in 2005, at which time China recognized India's annexation of Sikkim *(52)*.

Nepal and Bhutan kept out of direct participation in the conflicts between their bigger neighbours, but were indirectly affected to a certain extent *(52)*.

Burma, to which the British gave independence in 1948 (only three years after driving the Japanese out of it), had become by the 1960s a kind of new Tibet, deprived by its military rulers of contact with the outside world. Its Burmese name, Myanmar, is now widely used; and its capital, Rangoon, is called Yangon. A country long known as an exporter of rice and oil (petroleum) ran short of both, owing to misgovernment by the ruling junta that brutally suppressed all opposition. Perhaps overestimating their popularity, the military rulers (who called themselves the State Law and Order Restoration Council and later the State Peace and Development Council) permitted a free parliamentary election in 1990. When the

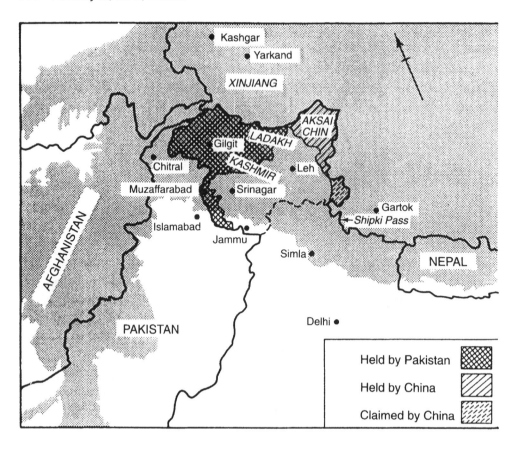

opposition won a decisive victory, the government refused to step down, prevented the parliament from meeting, and has kept opposition leader Aung San Suu Kyi under arrest since then.

In 2005, Burma's rulers suddenly announced that Pyinmana, an isolated mountain town, would replace Rangoon as the capital.

For 30 years there was almost continuous fighting between the junta's troops and some of the minority groups – mostly hill peoples in frontier areas. Between 1989 and 1995 ceasefire agreements were made with the Kachins, the Chins and other hill peoples, but in the south-east the Karens fought on until 2004. In the west, persecution of the Muslim community in Arakan had forced 200,000 of them to flee into Bangladesh; members of

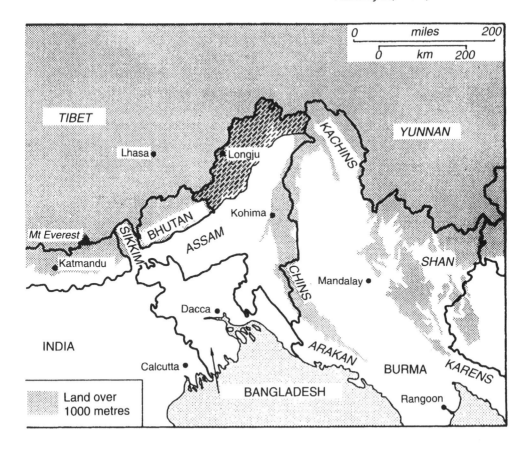

other groups fled into Thailand. Promised reforms have not materialized, and most minority groups have refused to disarm.

In the hill areas in eastern Burma formerly known as the Shan States there was a great increase in the cultivation of opium poppies, and by the early 1990s this region was growing 40% of the world's opium – much of which is processed into heroin. Starting in the late 1990s, the government sharply reduced the area under cultivation, but Burma remained the second-largest producer after Afghanistan *(50)*. Although the heroin was mostly smuggled out through Thailand or China, and some of the opium-producing areas were not under the Burmese junta's control, there were lively suspicions about the junta's connivance in drug production and trafficking.

54 China and Russia

In a long series of conquests, from 1580 to 1900, Russia and China took over all of northern Asia. Between them, they subdued the Mongolians and all the Turkic-language peoples, from the Azerbaijanis of the Caucasus to the Yakuts of eastern Siberia, and including the Kazakhs, Tatars, Uighurs and Uzbeks.

Then the two empires clashed. In 1858 and 1860 the Russians annexed China's Pacific-coast territory as far south as the border of Korea. In central Asia, between 1845 and 1895, they conquered the region between the Aral Sea and the frontiers of Iran and Afghanistan. They also asserted their power in Xinjiang (Sinkiang, or Chinese Turkistan), Mongolia and Manchuria, regions nominally ruled by a then weak China. In 1898 they acquired a naval

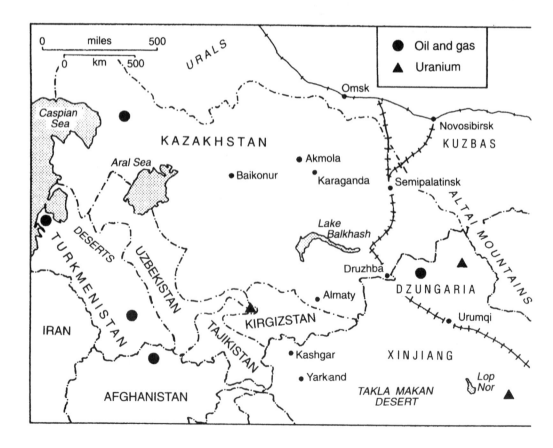

base in China itself, at Port Arthur (now part of the city of Lüda). Japan seized this base in 1905; the Russians regained it in 1945, but in 1955 they handed it back to China.

After Russia's 1917 revolutions the Turkic peoples of central Asia tried to regain their independence, but by 1924 they had been reconquered. A million Kazakhs fled into China, and millions of Russians and other Europeans were settled in the Soviet Union's central Asian regions. The Russians repelled China's attempts to regain control of Mongolia; after 1924 it was formally independent, but in reality Soviet-controlled (until 1990). The USSR annexed Tuvinia (Tannu-Tuva) in 1944.

After Japan's defeat in 1945, Soviet troops occupied north-east China (Manchuria) for a time. In the 1950s, however, Manchuria and Xinjiang ceased to be Soviet spheres of influence; China's new communist rulers took hold there. Millions of Chinese were settled in Xinjiang – until then peopled mainly by Uighurs and Kazakhs – and in Inner Mongolia (Nei Mongol).

When Sino-Soviet hostility developed in the 1960s, China began to complain that the Russians' nineteenth-century acquisitions of large border areas had been illegal. There were a series of armed clashes, notably at Damansky island in the Ussuri river, north of Vladivostok, and near Druzhba on the Kazakhstan–Xinjiang border. In 1974 the USSR, worried because its Trans-Siberian railway ran so close to the frontier, began to build the Baikal–Amur (BAM) line farther to the north. For both communist powers, central Asia

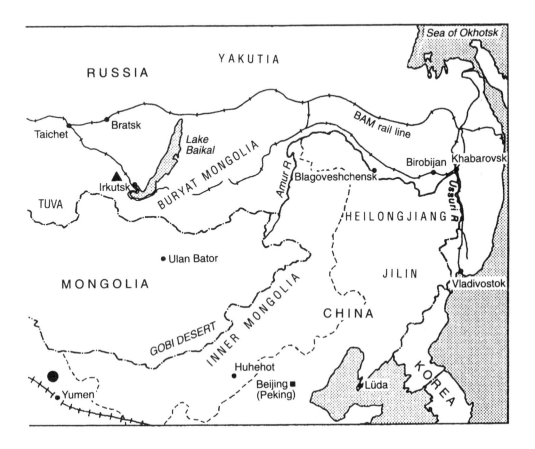

had become a site for nuclear weapon tests – China's at Lop Nor lake, the USSR's near Semipalatinsk and elsewhere in Kazakhstan – and a source of uranium.

In 1989 the USSR withdrew its troops from Afghanistan *(50)* and began to reduce its forces in Mongolia and on the Chinese frontier; this satisfied some of China's requirements for a 'normalizing' of relations. Tension was eased, and work was resumed on the completion of a rail link between Urumqi and the Soviet frontier at Druzhba. Soon, however, the disintegration of the Soviet Union and the emergence of independent states in central Asia led China's rulers to fear that Xinjiang would be affected. Chinese troops were flown to Kashgar in 1990 to put down what was officially called a Muslim 'rebellion'. The unrest spread to Urumqi, and reportedly involved Kazakhs and Tajiks as well as Uighurs.

The breakup of the USSR created a buffer zone in central Asia, and Russo-Chinese relations continued to improve. China's rulers were relieved when they saw old-style communists retaining or regaining control in central Asia's new states. But there was still unrest among the 8 million Uighurs of Xinjiang, encouraged by a clandestine 'Uighurstan Liberation Front' based in Kazakhstan (where about 250,000 Uighurs live). In 2001, China, Russia and several central Asian states formalized earlier agreements in the Shanghai Co-operation Organization *(20)*, among whose aims was the suppression of the 'evil forces' of separatism and terrorism.

In the early 1990s, massive sales of arms to China made up 2% of Russia's exports. Russia needed the money badly, but its generals grew nervous about China's new military strength. The last of their interminable border disputes – about sovereignty over islands in the Amur and other rivers marking the border between Siberia and Manchuria – was finally resolved by treaty in 2005. However, China's huge and growing population still made it an uncomfortable neighbour for Russia's thinly peopled Far Eastern territory, as Russia feared that Chinese immigration would change the region's ethnic composition. Meanwhile, both governments began to worry about the risk of a new conflict in the region being set off by North Korea *(59)*.

55 China and Other Neighbours

After long weakness, China is once more the strongest power in East Asia. It has the world's biggest population and its biggest army (with nuclear arms). Since its rulers lost faith in Marxism and began to shape a market economy, its gross domestic product (GDP) has grown by 8–10% a year and is now, by some measures, the world's second-biggest *(1, 2, 4)*.

In China's first years of communist rule, after 1949, it was allied to the Soviet Union; its army intervened in Korea; it supported communist forces fighting in Indochina and (less successfully) in the Philippines and Malaya. When China broke with the USSR and got on better terms with the United States, its non-communist neighbours became less anxious about its intentions. The only neighbour it has fought with during the past forty years is communist Vietnam *(61)*. It did not try to take over Hong Kong before the set date, 1997; it set no date for the reunion with Taiwan which, it insists, will come some day.

In recent years, those actions that have worried China's neighbours have revealed national, not ideological, motives *(6)*; it has shown more interest in offshore oil than in encouraging revolutions. In the 1970s and 1980s it drove Vietnam's forces out of the Paracel islands and attacked them in the Spratlys. In the early 1990s the Philippines and Malaysia, which were maintaining small garrisons on some of the Spratly islets, found China deaf to their protests about its encroachment and their appeals for negotiation; but it left Taiwan's garrison there undisturbed. As with similar disputes elsewhere in Asia *(56, 58)*, sovereignty over uninhabited islets brings control of potentially oil-rich seabed, but exploitation cannot proceed without boundary demarcation as in the North Sea in Europe *(5, 22)*.

By the 1980s, South Korea, Taiwan, Hong Kong and Singapore were being called 'the four little dragons' because of their rapid economic progress and their close relationship to China. (Korea's culture owes much to China's; the populations of Taiwan and Hong Kong are almost wholly, that of Singapore predominantly, Chinese.) Those four were also often grouped with South-East Asia's other 'newly industrialized countries' (NICs) under the label of 'little tigers' *(2, 60)*.

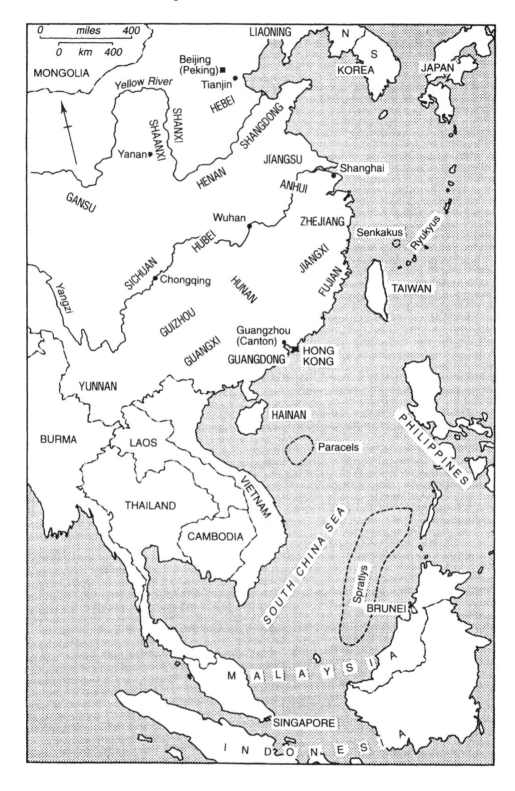

56 Taiwan

After the communists' victory in the 1946–9 Chinese civil war, the defeated Kuomintang (Nationalist) government took refuge in Taiwan (Formosa). Although ruled by Japan from 1895 to 1945, Taiwan was an old Chinese province. For many years the refugees from the mainland, who formed less than 20% of the population (now 23 million), kept power in their own hands and insisted that the true government of China was the one in Taiwan, which would in time return to Beijing (Peking).

During the 1970s the mainland government took over China's seat at the United Nations and was recognized by the United States. China refuses to establish formal diplomatic relations with countries that recognize the independence of 'Taiwan Province', and by 2002 only twenty-seven countries maintained such ties with the 'Republic of China' on Taiwan. But most others kept informal links with it, because it had become a major trading nation. By the early 1990s it was the world's tenth-biggest exporter, with huge foreign-exchange reserves and an annual GDP growth rate of over 7%. The rate has since slowed, but its per capita GDP is now higher than that of many European Union members.

The Kuomintang ruled Taiwan until 2000, but the first fully democratic election was held in 1992. The main opposition party had proposed that, if a referendum showed enough support, the island should declare its separation from China. This angered China, which renewed former threats to invade Taiwan, and added pressure in 1995–6 by staging army exercises along its coast and test-firing missiles into the sea near the island. The United States then sent warships to the area; in the 1950s it had been committed to protecting Taiwan, and it still sought to deter China from contemplating an invasion. China did not go so far as to resume shelling Quemoy and Matsu, the Taiwan-held islands near its coast, which it had left in peace since the 1970s. Later in 1996, Taiwan sided with China in a dispute with Japan over the uninhabited Senkaku (or Diaoyu) islands; there were hopes of finding offshore oil near them *(55)*.

Despite its statements favouring separation, or at least open discussion of the possibility of independence, the opposition has not risked formal moves towards independence since its victory in the 2000 presidential election. Ever closer economic ties between Taiwan and China have continued to be balanced by a tense political relationship. In theory, trade between them was banned until 2001; in fact, Taiwan traded with the mainland (mostly via Hong Kong) before then. China (including Hong Kong) is now Taiwan's largest trading partner and overseas investment destination. Both joined the World Trade Organization in 2001–2 (Taiwan as 'Chinese Taipei'), and 2005 saw the first direct flights since 1949.

The political relationship is one of brinkmanship tied to the understanding that war is in neither side's economic interest. Since the 1990s, rapid economic growth in China has funded an unprecedented military buildup there. China threatens invasion if Taiwan

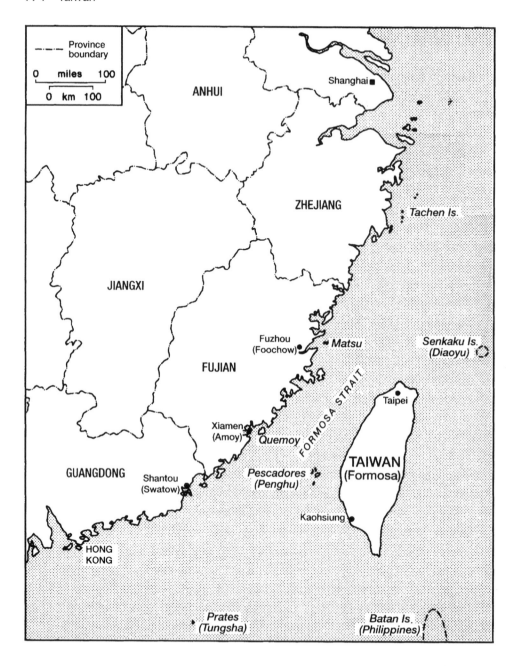

declares independence. Taiwan has bought advanced weapons and missile defence systems from the US – which discourages independence while still promising to defend it against invasion – but so far has avoided the formal steps towards sovereignty that could provoke Chinese military action. China's policy towards Hong Kong since the 1997 takeover *(57)* is of interest to Taiwan as an indicator of whether there is any prospect of Taiwan being eventually reunited with the mainland on acceptable terms – highly unlikely without democracy in China.

57 Hong Kong and Macau

In the nineteenth century China's coast became speckled with foreign-ruled enclaves – British, French, German, Japanese, Portuguese and Russian. After 1955, when China regained Port Arthur from the Russians *(54)*, only Hong Kong and Macau (Macao) remained.

The island of Hong Kong (Xianggang) was ceded to Britain in 1842, and the Kowloon area in 1860 (after the first and second opium wars respectively); in 1898 the New Territories were acquired on a lease running to 1997. The colony's population, which was only 700,000 after the 1941–5 Japanese occupation, rose to 2.5 million by 1955, largely because of a flood of refugees from China. By 2006, swollen by further immigration, it was 7 million (and 95% Chinese). Income per head was far above the level on the mainland and among the highest in the world. Manufacturing, trading and financial activity had turned Hong Kong into one of East Asia's prosperous 'little dragons'.

However, it faced the prospect of being taken over by China in 1997. After the expiry of the New Territories lease, what was left would be neither defensible nor viable. In 1984, Britain and China signed an agreement for the transfer in 1997 of the whole of Hong Kong to China, which undertook to allow it a degree of autonomy for 50 years – the 'one country, two systems' plan.

In 1987, Portugal and China made a similar agreement for a handing over of Macau in 1999. In 1974, Macau (population 500,000, also 95% Chinese) had become, formally, a Portuguese-administered area of China; in practice, the Chinese already controlled it.

Both transfers occurred peacefully. China clearly appreciated the value of Hong Kong's thriving economy as a source of investment and a model for mainland cities. In 1979, China created 'special economic zones' (SEZs) around Shenzhen and Zhuhai, adjacent to Hong Kong and Macau, and gave foreign firms incentives to operate there, producing mainly for export. Hong Kong firms run by 'overseas Chinese' invested massively in China as soon as it permitted foreign investment; and China had invested in Hong Kong – which also suggested that it would seek to preserve Hong Kong's prosperity. That prosperity, however, had been created with minimal state interference. Although by 1997 China's communist government had for twenty years been steadily transforming the mainland's Marxist-style economy into a market economy, there were still fears that it might not resist the urge to interfere in Hong Kong's.

This did not happen, and in fact China's economic boom has been led by its regions and cities that sought to follow the Hong Kong model. In the 1980s and 1990s, more SEZs were created along the coast, including the entire island of Hainan. By the late 1990s, a wide income gap had opened between the better-off coastal provinces and the poorer inland. Movement of tens of millions of people from rural areas and the interior to cities and

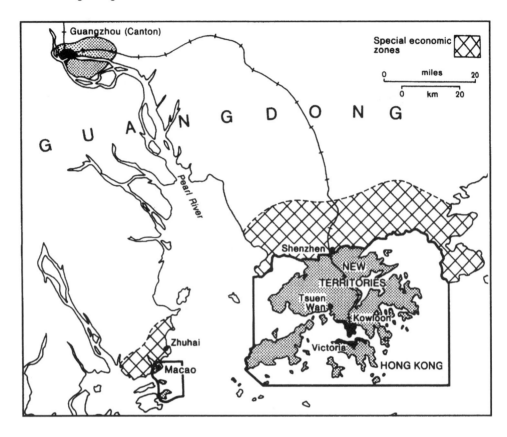

coastal provinces occurred despite official restrictions on internal migration. This boosted the already large urban populations in Beijing, Shanghai and the Pearl river delta from Hong Kong to Guangzhou (Canton) – where millions of people work in factories owned by Hong Kong and foreign investors.

It was clear even before 1997 that China would not wait for 50 years without interfering in Hong Kong politics. Unlike other British dependencies, Hong Kong had never had much chance to learn to operate democratic institutions. Only after 1990 were the electorate and the legislature's powers enlarged (and not widely, even then). But this late and limited move towards democracy was still unacceptable to China, which named its own appointees to replace the elected legislators, and indicated that it would not tolerate political opposition. At the same time, faced with large popular demonstrations, China has not made major changes to the existing system, and Hong Kong retains more press and religious freedom than the mainland.

58 Japan

Defeat in 1945 stripped Japan of an empire, acquired in fifty years of conquest, which at its peak embraced most of East Asia. After centuries of self-imposed isolation, Japan had adopted a policy of territorial expansion in 1895. By 1910 it had annexed Korea and taken Taiwan from China and southern Sakhalin (Karafuto) from Russia. After the 1914–18 war it took over, under a League of Nations mandate, the Caroline, Mariana and Marshall islands in Micronesia, which had been German colonies. In 1931, taking advantage of China's lack of any strong central government, the Japanese occupied its north-eastern region, Manchuria, where in 1932 they established a puppet state called Manchukuo. During the 1930s they conquered a large part of eastern China.

In 1940 they entered French Indochina (now Vietnam, Laos and Cambodia). In 1941 they occupied Thailand, and by mid-1942, after attacking all the American, British and Dutch territories in the region, the Japanese were in control of the Philippines, Guam and Wake Island; Hong Kong, Malaya, Singapore, British Borneo and most of Burma; all the Dutch East Indies (now Indonesia); much of New Guinea, most of the Solomons and all the Gilbert Islands; and, far to the north, the western Aleutians.

All these conquests were wiped out in 1945. China recovered Taiwan. The Soviet Union recovered southern Sakhalin and seized the Kurile islands; its forces also occupied Manchuria and North Korea, staying until Chinese and Korean communists could take over. American forces occupied South Korea, as well as Japan itself, where Commonwealth contingents joined them.

In the peace treaty concluded in 1951 with all the victorious allies except China and the Soviet Union, Japan renounced its claims on Taiwan, Korea, Sakhalin and the Micronesian islands. The occupation of Japan was ended; Japan concluded a mutual security treaty with the United States and gave it the right to keep forces in Japan for joint defence.

The Micronesian islands became a trust territory under American administration *(66)*. Between 1953 and 1972 the Americans returned to Japanese control, group by group, the Bonins, the adjacent Volcano Islands (including Iwo Jima) and the Ryukyus (including Okinawa, where the US retained rights to a base).

In 1978, Japan was at last able to conclude a peace treaty with China. It had still signed no peace treaty with the Soviet Union, largely because of a continuing dispute over the small Habomai, Shikotan, Kunashiri and Etorofu (Iturup) islands, which lie near Hokkaido, the northernmost of Japan's main islands. The USSR had seized these islands in 1945, along with the rest of the Kuriles chain, and it (and later independent Russia) rebuffed a long series of Japanese requests for their return. Japan also claims the South Korean-controlled Dokdo (or Takeshima) group in the Sea of Japan, midway between the two

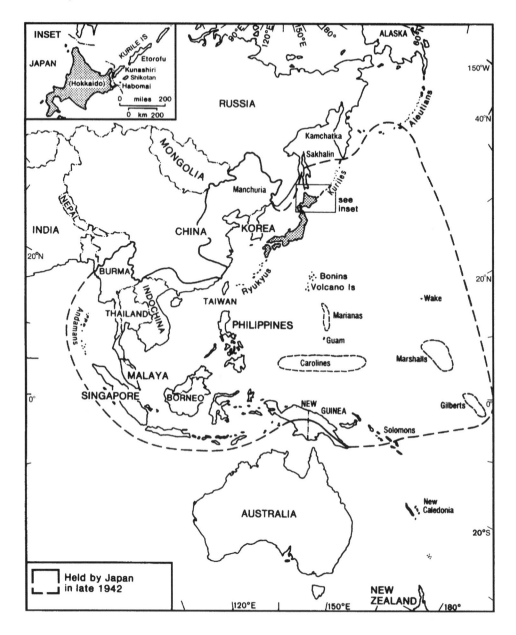

countries. The islands are uninhabited but bring control of potential undersea gas or oil fields *(5)*.

Japan's brief but spectacular phase of territorial conquest had been spurred on by the belief that a populous island nation, with very few mineral resources, needed to acquire overseas 'living space' and sources of fuel and raw materials. It was the oil, rubber and mineral resources of South-East Asia that Japanese armed forces went south to seize in 1941–2. Yet it was after 1945, when Japan was forced back to its pre-1895 limits, that it developed its economy so successfully that it became a superpower in economic terms *(2)*.

It still depends heavily on oil imported mainly from the Middle East and on imports of minerals from many countries, including Australia, Canada, India and South Africa. But its industrial strength and its importance as a market for countries that export raw materials have gained it a position of great influence in the Asian and Pacific regions, and maintained it in that position far longer than its armies ever did. And, after failing to defeat Americans and Europeans in war, it bettered them in peace – building up huge surpluses in its trade with them, making large investments and establishing many factories in their countries as well as in Asian ones, and coming to play a powerful role in international finance. After 1985, Japan became the world's biggest creditor nation, its surpluses in visible trade being augmented by the huge amounts it received annually in dividends and interest.

The decades of expansion ended in the early 1990s with the collapse of the 'bubble economy', in whose heyday the city of Tokyo had been worth more on paper than the entire United States. In the next decade, banks failed from the burden of bad debts, unemployment rose, and property prices and the stock market fell by more than half. Despite its severity and long duration, the crisis was domestic: Japan maintained its foreign surplus throughout. The rapid increase in Chinese exports has, however, diverted attention from Japan, and Japan's ageing and declining population puts its future economic growth into question. Unlike most other rich countries, Japan has never welcomed foreign migration, preferring automation – Japan has by far the world's largest stock of industrial robots – to immigration. In a population of 128 million, only 1% are not ethnic Japanese.

In its 1946 constitution Japan renounced the use of force as a means of settling international disputes. The principle of pacifism has been modified since 1956 by its creation of 'self-defence forces', and by the 1990s its defence budget was one of the biggest in the world. It has, however, continued to regard its pact with the United States as the main guarantee of its security; and this has been seen as an important factor making for stability in the East Asian region. American forces stationed in Japan are about 47,000 strong. In 1996, after protests about their occupation of large areas in Okinawa, it was agreed that some of this land would be relinquished and one of the American bases closed.

59 Korea

In Korea, an old kingdom where Chinese cultural influence was strong, a struggle for control between Russia and Japan ended when Japan seized the country in 1905 *(58)*. After Japan's defeat in 1945, American forces occupied the south of Korea and Soviet forces the north; the dividing line was the '38th parallel' (38° North). The USSR then installed a communist regime at Pyongyang in the north; in the south, where two out of three Koreans live (today there are 50 million in the south, 23 million in the north), an elected government was set up in the old capital, Seoul.

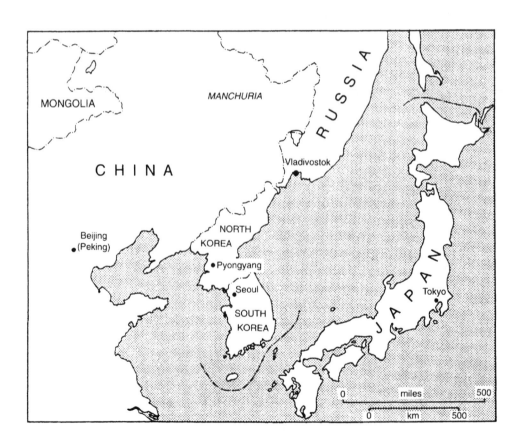

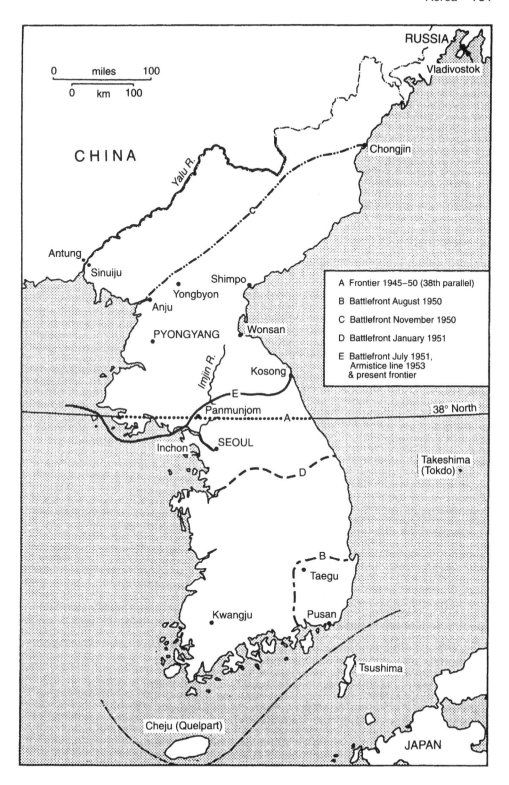

RUSSIA

Vladivostok

CHINA

Yalu R.

Chongjin

Antung

Sinuiju

Shimpo

Yongbyon

Anju

PYONGYANG

Wonsan

Imjin R.

Kosong

E

Panmunjom

A

38° North

SEOUL

Inchon

Takeshima
(Tokdo)

D

B

Taegu

Kwangju

Pusan

Tsushima

Cheju (Quelpart)

JAPAN

A Frontier 1945–50 (38th parallel)

B Battlefront August 1950

C Battlefront November 1950

D Battlefront January 1951

E Battlefront July 1951,
   Armistice line 1953
   & present frontier

0   miles   100

0   km   100

The American and Soviet forces withdrew; but North Korea was heavily armed by the USSR, and in 1950 it invaded the south, quickly capturing Seoul. American troops were sent to help resist the invasion; the United Nations urged other nations to help, and eventually sixteen nations would send forces to fight in Korea under the UN flag; but the first few units to arrive were driven back, together with South Korea's own small army, into the area around Pusan.

A new American force was landed at Inchon, cutting the invaders' supply lines; the North Koreans were driven back across the 38th parallel but refused to make peace. The UN forces advanced northwards. When they approached China's frontier China sent a large army into Korea, which drove south until Seoul once more fell into communist hands. But by June 1951 the UN allies had advanced across the parallel again, and China agreed to start talks. The talks, mostly held at Panmunjom, dragged on until 1953, when (after Stalin's death) an armistice was signed.

A 1954 peace conference ended in deadlock, and Korea remained divided along the armistice line. The North's military pressure led South Korea, too, to raise large armed forces, which dominated its politics, one general after another taking over the presidency. However, South Korea prospered so much that, in only thirty years, income per head soared from a low level like India's to one more like that of Greece. Impatience with military domination led to a revolt at Kwangju in 1980, to reforms in 1987, and to the election of a civilian president in 1993.

North Korea's unpredictable communist rulers kept it sealed off from the outside world. Their policies (which included keeping more than a million soldiers under arms) made it very poor, and it became even poorer when Soviet and Chinese subsidies were withdrawn. There were fears that its rulers, out of sheer desperation, might attempt to invade the south – which, though protected by a large army and 30,000 US troops, has its capital within 30 miles of the border. Severe famines beginning in 1996 killed perhaps 2 million; the number would have been higher without large aid shipments from South Korea, China, the UN and the US. The North Korean government also stands accused of exporting heroin, producing high-quality counterfeit 'superdollars' and acquiring nuclear technology from Pakistan.

Fears that North Korea might try to produce nuclear weapons *(4)* increased in the 1990s. To induce it to shut down its plutonium-producing installations at Yongbyon, north of Pyongyang, it was offered, in 1994–5, American, Japanese and South Korean aid in the form of safer nuclear power plants and, until those plants could be built, supplies of oil. In 1998, North Korea test-fired a missile over Japan into the Pacific, and in 2002 the US accused North Korea of maintaining a secret uranium-enrichment programme. Acknowledging this, North Korea ejected international inspectors and restarted the Yongbyon reactor – cancelling the 1994 agreement – and in 2005 claimed to have nuclear weapons. Negotiations involving both Koreas, China, Japan, Russia and the US initially led to an agreement by the North to give up its nuclear programme but later collapsed.

60 South-East Asia

Before the 1939–45 war the only independent nation in South-East Asia was Thailand (then called Siam). The rest of the region was under European or American rule: French in Indochina (Vietnam, Laos and Cambodia); Dutch in the East Indies (now Indonesia); British in Burma, Malaya and northern Borneo; American in the Philippines. The Japanese moved into Indochina in 1940 and overran the whole region in 1941–2 *(58)*. In 1945 the Dutch were unable to regain full control of Indonesia, which later became independent *(64)*. The United States gave independence to the Philippines in 1946; Britain gave it to Burma in 1948 and to Malaya in 1957 *(53, 63, 67)*. In Vietnam the French faced a communist-led independence movement which, for some time, had the support of both the USSR and China *(61)*.

By 1954, North Vietnam was being taken over by a communist regime which was also gaining footholds in Laos, Cambodia and South Vietnam. Communist guerrillas were active in Burma, Malaya and the Philippines, and there were fears about the new strong China's influence on the region's large Chinese communities (about 30 million Chinese live in South-East Asia, and their economic importance is greater than their numbers would suggest). Thailand and the Philippines joined Australia, Britain, France, New Zealand, Pakistan and the United States in signing the 1954 Manila treaty on South-East Asian defence, often called SEATO. These allies agreed to act together against any attack in the region on one of them, or on Cambodia, Laos or South Vietnam (although any allied action in one of those three states would require its consent).

No joint defence action was ever taken under SEATO's formal authority, although American, Australian, New Zealand, Philippine and Thai forces went to fight in Vietnam in the 1960s *(61)*. In the 1970s, SEATO's activities were ended. By then, much had changed in the region. In Indochina, military victory had gone to the communists; but Vietnam's communist rulers had broken with China, siding with the USSR against it and initiating a new quarrel over Cambodia *(62)*. Meanwhile a new grouping, more truly regional than SEATO, had been formed.

The Association of South-East Asian Nations (ASEAN) was formed in 1967 by Indonesia, Malaysia, the Philippines, Singapore and Thailand; Brunei joined them in 1984, Vietnam in 1995, Laos and Myanmar (Burma) in 1997, and Cambodia in 1999. ASEAN was never a military alliance, but its members sought to unite in pursuing their common interests – for example, in trade negotiations with the European Union. In 1994 the ASEAN Regional Forum (ARF) began a series of consultative meetings in which China, the EU, Japan, Russia and the United States all took part. The ASEAN members also joined the Asia–Pacific Economic Co-operation (APEC) forum *(2)* and have ties with north-east Asia through 'ASEAN+3' meetings that include Japan, South Korea and China. In seeking to extend its regional influence, China also supported an East Asian summit that was held in 2005. This

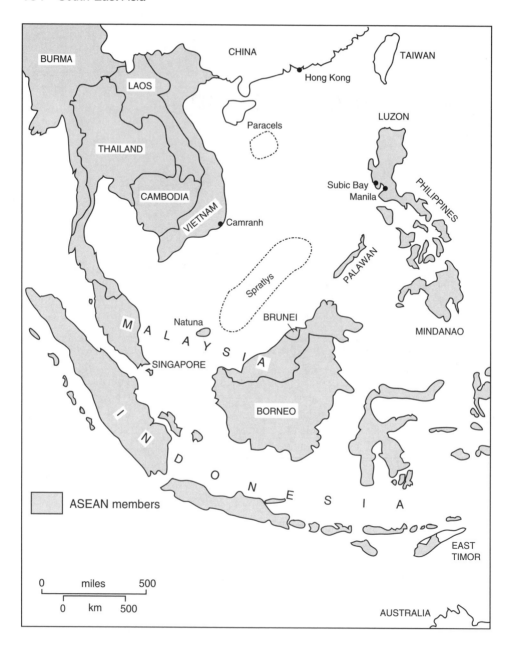

excluded the US but – because other participants were wary of Chinese influence – brought in India, Australia and New Zealand.

The big American naval and air bases in the *Philippines* – at Subic Bay and Clark Field, near Manila – were relinquished in 1991–2 (when the breakup of the USSR led to the abandoning of the Soviet naval base at Cam Ranh Bay in Vietnam *[61]*). The Philippines had been troubled since 1969 by insurgency among the Moros, its Muslim minority that live mostly in Mindanao and other southern islands. In 1993 the Moro National Liberation Front (MNLF, the main rebel group) agreed to negotiate with the government, and in 1996 a peace pact, providing for more local autonomy, was signed. Another group, the Moro Islamic Liberation Front, continued its activities; ceasefires have ended in violence, but talks began again in 2005. The US strengthened military links following the 9/11 terrorist attack *(8)* in order to pursue the Abu Sayyaf, an offshoot of the MNLF linked to al-Qaeda that seeks an Islamic state in the southern Philippines.

Partnership in ASEAN has helped its members to avoid clashes over some territorial disputes, such as the Philippines' old claim to Sabah in Malaysia. A dispute between Indonesia and Malaysia over Ligitan and Sipadan, two small islands east of Borneo, was taken to the International Court at The Hague (which settled in Malaysia's favour in 2002), and Malaysia and Singapore brought a similar problem to the Court in 2003. In the South China Sea, although Brunei, Malaysia, the Philippines and Vietnam all claimed some of the Spratly islets (hoping to find offshore oil near them), there was little friction between them; they were, instead, united in protesting against China's intrusions and its sweeping claim to the whole island group *(55)*.

A problem for the whole region emerged when millions of people began to try to escape from Indochina after the communist conquests of 1975. Cambodian refugees mostly fled into Thailand, but many of the far more numerous refugees from Vietnam put out to sea in small craft. Those 'boat people' who survived these voyages found only a temporary refuge in the ASEAN countries, which demanded assurances that they would be resettled elsewhere. The 200,000 'boat people' who reached Hong Kong presented particularly acute problems for a crowded little territory which was already struggling to avoid being swamped by Chinese seeking to escape from the mainland.

The United States accepted more than a million refugees from Vietnam; Australia, Canada and France about 130,000 each; smaller numbers went to a dozen other countries. Some 50,000 were induced to return from Hong Kong to Vietnam. Economic reforms improved the situation in Vietnam after 1986, and the Hong Kong refugee camp closed in 2000.

61 Indochina

Cambodia, Laos and Vietnam all came under French rule between 1860 and 1900, forming French Indochina. During the 1940–5 Japanese occupation, the communist-led Vietminh movement launched a resistance campaign in northern Vietnam (Tonkin, or Tongking). In 1945 it set up a government in Hanoi. Although the French reoccupied Hanoi in 1946, they faced a long conflict with the Vietminh. In 1954, after the Vietminh had trapped a French force at Dien Bien Phu, ceasefire agreements were signed. Vietnam was divided at the 17° North parallel. The French withdrew from North Vietnam, where a communist government was again installed in Hanoi. In South Vietnam the French completed the transfer of sovereignty to the government in Saigon. About 800,000 Vietnamese fled from north to south.

France also withdrew its forces from Cambodia and Laos, whose governments now had full sovereignty. In Laos attempts to include the communists in a coalition government failed, clashes multiplied, and by 1961 communist troops were threatening Vientiane, the capital. A new coalition was installed in 1962, but the conflict was soon resumed. Weak governments in Laos and Cambodia allowed North Vietnam to move its troops across their territory (along the 'Ho Chi Minh trail', which in reality consisted of many tracks through the hills) so that they could infiltrate South Vietnam from the west.

By the early 1960s, South Vietnam's government was losing control of many areas to 'Viet Cong' guerrillas armed, reinforced and directed from North Vietnam, which in turn was being aided by China and the USSR. The Americans, who were supporting South Vietnam in the hope of checking the southward advance of communist power, became more deeply involved in the struggle. By 1963 there were 16,000 US military personnel in Vietnam; in 1965, US aircraft began to bomb North Vietnam, and US ground forces arrived in the south; by 1968 there were 500,000 Americans fighting there, and alongside them and the South Vietnamese troops were contingents from Australia, New Zealand, the Philippines, South Korea and Thailand.

A series of peace offers failed; the Hanoi government was bent on, and eventually succeeded in, taking over South Vietnam. In 1969 the withdrawal of the US and allied troops began; it was completed by 1973. A ceasefire was then announced, but the conflict was soon resumed. By 1975 the North Vietnamese army fighting in the south was larger than South Vietnam's; it captured Saigon and the other cities, and Vietnam was forcibly united under communist rule. (Hanoi remained the capital; Saigon was renamed Ho Chi Minh City.) There followed a new flight of about 1,600,000 refugees. 'Boat people' escaped in small craft; thousands of them perished at sea – many at the hands of pirates – before they could reach land in places ranging from Hong Kong to Indonesia *(60)*.

In 1975, Laos was also taken over by communist forces which were effectively controlled

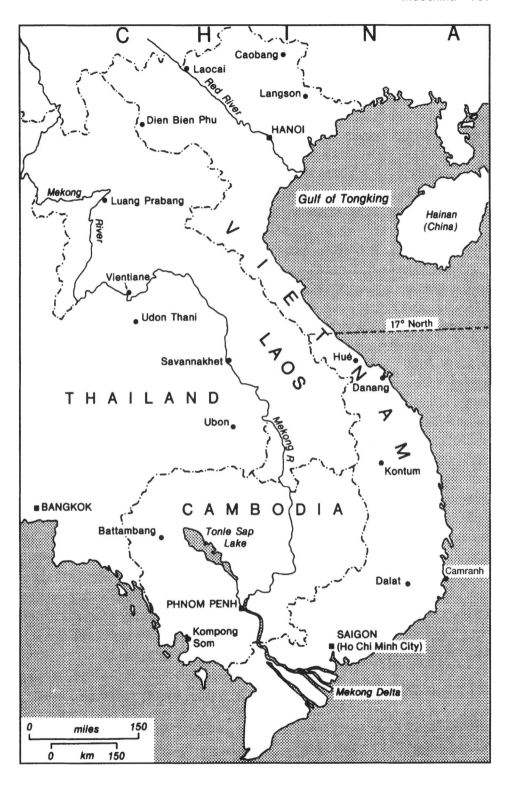

from Hanoi. However, the 'Khmers Rouges' who simultaneously took over Cambodia looked to China for support and were soon in dispute with the Soviet-backed Hanoi government *(62)*. When Vietnam invaded Cambodia late in 1978, China responded by staging a limited invasion of Vietnam early in 1979 – but, after capturing some border towns, it withdrew its troops, which had suffered heavy losses. Thus, in three decades, Vietnam had turned away three of the post-1945 great powers.

For a time, the whole of Indochina became a Soviet sphere of influence. The Soviet navy secured a base at Cam Ranh Bay which increased its ability to operate in the Indian Ocean as well as in Far East waters *(42)*. By the late 1980s, however, the USSR began to regard Indochina as more of a liability than an asset. It reduced its subsidies to Vietnam's rulers and withdrew its forces from Cam Ranh Bay. Those rulers, alarmed by the changes they saw in the Soviet sphere, sought to improve their relations with China. When China offered them little help they turned elsewhere, opening up the economy enough to obtain investments and aid from Japan and other countries. In 1995, Vietnam joined ASEAN *(60)* and established diplomatic relations with the United States; economic ties grew rapidly in the next decade. Though still a one-party state under tight communist control, Vietnam had chosen the Chinese path towards export-oriented growth.

62 Cambodia

Cambodia (Kampuchea) is the remnant of the old Khmer kingdom which once included the Mekong delta. It was a French protectorate from 1863 until 1953. Unlike Laos, it has no border mountains separating it from Vietnam; some Vietnamese settled in Cambodia, although there was an old antagonism between the two neighbours. During the 1960s, Cambodia tried to keep out of the Vietnamese conflict, turning a blind eye to the way North Vietnamese troops were sent along the 'Ho Chi Minh trail' through Laos and eastern Cambodia to fight in South Vietnam *(61)*; but in 1970 Cambodia itself became a battlefield. Its communist guerrillas, called the Khmers Rouges ('Red Khmers'), got the upper hand

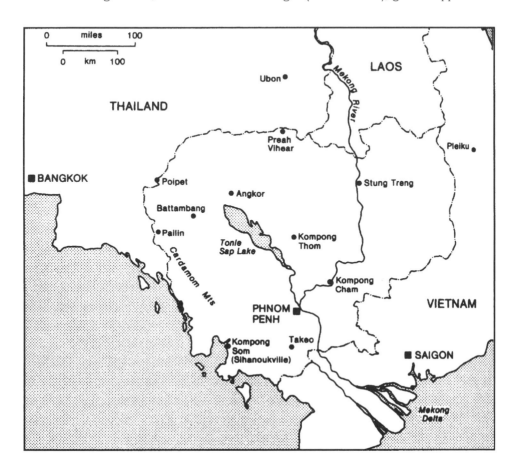

when American support for the government was withdrawn. In 1975 they captured Phnom Penh, the capital, and took over the country.

Cambodia's communist rulers became infamous for doctrinaire ruthlessness. They forcibly emptied the cities, created a famine, and killed at least 1.5 million people; thousands fled into Thailand. While Vietnam toed the Soviet line, the Khmers Rouges sided with China. In 1978, after many border clashes, communist Vietnam invaded communist Cambodia. A puppet government (headed by dissident Khmers Rouges) was installed in Phnom Penh, backed by 150,000 Vietnamese troops. More Cambodians fled to Thailand; others got only as far as refugee camps along the Thai border, which were mostly controlled by Khmer Rouge guerrillas.

In 1982 three resistance movements, one of them run by Khmers Rouges, formed an anti-Vietnam coalition government which, although it held only a few border areas, was recognized by many other countries. By the late 1980s, Vietnam was no longer receiving large Soviet subsidies, and it could not maintain its large army of occupation in Cambodia, which it withdrew in 1989. A new struggle among the local factions ensued, and the UN sent a force to restore order and supervise an election, held in 1993. Although the Khmers Rouges boycotted it (and refused to disband their forces), there was a 90% turnout. The 370,000 Cambodians who had fled to Thailand began to return home.

Following the election, a government was formed in which power was divided – uneasily – between the main non-communist (royalist) party and the party that had formerly collaborated with the occupying Vietnamese forces. Since then, this coalition has kept power with a mix of electoral success and force.

In the mid-1990s, Khmer Rouge forces still held areas in the north and west, mostly close to the Thai frontier. In 1996 a split developed among their commanders; some of them announced that, if they were allowed to play a part in Cambodia's politics, they would stop fighting. Government troops were sent to support these dissidents in their clashes with the remaining diehard Khmers Rouges. In 1997 the leader of the ex-socialist party (formerly allied with Vietnam) mounted a coup against the prime minister, a member of the royal family. This slowed Cambodia's entry to ASEAN *(60)*, but the coalition returned to power after another election.

63 Malaysia and Singapore

Malaysia, a monarchical federation, consists of the sultanates and states of mainland Malaya (including Penang and Malacca) and the Borneo states of Sabah (formerly British North Borneo) and Sarawak. Four-fifths of its 26 million inhabitants live on the mainland. About half are Malays – who are mostly Muslims – but a quarter are Chinese and 7% Indians; in the Borneo states indigenous groups form a majority of the population.

Mainland Malaya became independent in 1957, after a long period of British rule interrupted by the Japanese occupation of 1942–5. From the 1950s onwards, democratic elections produced governments based on a Malay-led alliance that included Chinese and Indian parties. The main opposition came from an Islamic party supported by Malays who resented the economic strength of the Chinese. A series of steps were taken to increase the Malays' share of the economy, and to meet demands for the upholding of Muslim traditions; but the alliance-based governments resisted the more extreme demands of Islamic fundamentalists. Whether pro-Malay policies should continue has come into question as the economy as a whole and the Malay-controlled share have both increased.

The decision by Sabah and Sarawak to join Malaysia in 1963 was resented by Indonesia, which launched raids across the frontier. These attacks were repelled with the help of British forces, and in 1965 Indonesia abandoned its policy of 'confrontation'.

Singapore, an island territory with an original area of 225 square miles (expanded by a fifth through land reclamation), developed under British rule into a major port. It now has 4 million inhabitants, 76% of them Chinese, 14% Malay and 9% Indian. In 1963 it joined the newly enlarged Malaysian federation; but mainland Malays were unhappy about taking in such a large Chinese element, and in 1965 Singapore had to withdraw from Malaysia. As an independent republic, Singapore further developed its role as a trade and finance centre, and adapted itself to the new high-technology era in industry so successfully that by 2005 its income per head nearly matched that of Japan.

Malaysia's economy grew rapidly in the 1990s as foreign investment in manufacturing led to electronics replacing rubber and minerals as the most important exports. The Asian financial crisis of 1997 and increasing competition from lower-wage China slowed growth. Malaysia sought to move from manufacture to software and technology production through creation of a Singapore-sized 'multimedia super corridor' adjacent to the capital, Kuala Lumpur.

The sultanate of *Brunei* (population 400,000, two-thirds Malay) remained under British protection until 1984, when it became fully independent, with the formal name of Brunei Darussalam. It was formerly distinguished by its possession of a rich oilfield, but offshore finds have now made Malaysia a bigger producer of oil.

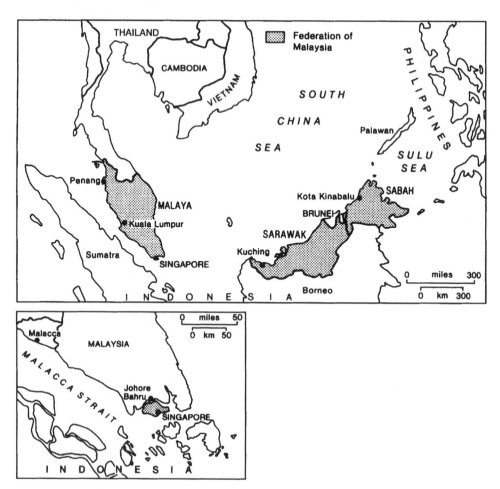

Malaysia and Singapore were founder members of the Association of South-East Asian Nations, and Brunei joined ASEAN in 1984 *(60)*.

64 Indonesia and New Guinea

A huge archipelago stretching 3,000 miles from east to west, *Indonesia* occupies the territory that was formerly the Dutch East Indies. Japan seized the territory in 1942; and Holland, which had itself been invaded and occupied by Nazi Germany, was unable to restore its authority after Japan's defeat in 1945. Fighting between Indonesians and Dutch was halted with the help of a United Nations observer mission, and sovereignty was transferred to an independent Indonesian government in 1949. The capital, Batavia, was renamed Jakarta.

The western half of New Guinea (then called West Irian, now known as Papua or Irian Jaya), whose Melanesian peoples had nothing in common with the Malay-speaking, largely Muslim Indonesians, remained in Dutch hands after 1949; but Indonesia demanded this territory, asserting its right to take over the whole of the former Dutch East Indies. After attacks by Indonesian forces, a transitional regime backed by a UN force was installed there in 1962. Indonesia took over Papua in 1963, and later transferred many settlers to it from other islands.

Three-fifths of Indonesia's population of 240 million still live in crowded Java, although a 'transmigration' programme has moved several million Javanese to other islands. Besides encouraging rainforest clearing, this has shifted the ethnic and religious balance of areas previously dominated by non-Malay peoples, practitioners of traditional religions, Christians or Hindus. One local problem has concerned the Christians of Ambon (Amboina) and nearby South Moluccan islands; thousands of them emigrated to Holland after 1949, and some continued to demand separate independence for the South Moluccas.

A more widespread problem concerns Indonesia's 8 million Chinese. At times, this scattered minority has been victimized in outbreaks of mob violence. In the early years of independence Indonesia's rulers aligned their foreign policy with that of communist China; but there was a sharp break in 1965, when Indonesia's army seized power, claiming that it had forestalled a communist coup which the Chinese minority was backing. Many of the Chinese were jailed, and many killed. Diplomatic relations with China were broken off, and were not resumed until 1990.

Before 1965, Indonesia had resorted to mounting raids across its border with Sabah and Sarawak in an attempt to prevent them from joining the Malaysian federation *(63)*. In 1975 it seized *East Timor* when Portugal withdrew from that territory, and fighting broke out between groups who were seeking independence and others who wanted to join Indonesia. The Indonesian occupying forces met with continued resistance, which they suppressed so brutally that the UN refused to recognize the annexation of East Timor (renamed Loro Sae).

Economic crisis in 1997 and popular protest led to the 1998 resignation of President

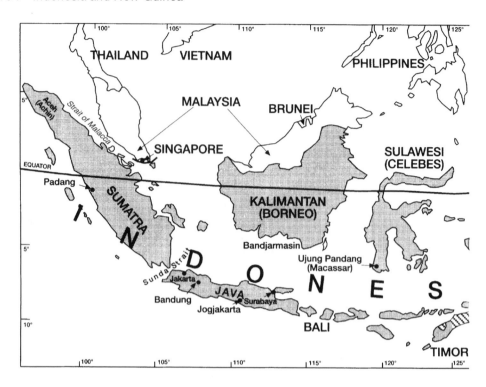

Suharto, only the second Indonesian ruler since independence. Democratic elections were held in 1999, and East Timor was allowed to vote on independence the same year. Following further Indonesian military intervention and the establishment of a UN peacekeeping force, full independence was gained in 2002. An agreement with Australia promised substantial future revenue from oil and gas fields in the Timor Sea.

With continued transition to democratic government since 2000, the Indonesian government became more willing to negotiate with separatist movements and offer regional autonomy. Aceh, at the northern end of Sumatra, fought Dutch control before the Second World War and objected to incorporation into Indonesia and to government-supported immigration of Javanese. A separatist movement challenged Indonesian control from 1976. A peace agreement signed in 2002 failed; but, following the devastation of the tsunami centred off Sumatra that killed 200,000 people around the Indian Ocean in 2004, Indonesian and Acehnese leaders were able to agree on supporting reconstruction efforts, increasing regional autonomy, and – perhaps most important for lasting peace – withdrawing the Indonesian army.

Terrorist attacks in Bali and Jakarta since 2002 and the threat of piracy in the Strait of Malacca between Sumatra and peninsular Malaysia have increased foreign interest in one of the world's busiest sealanes, the main route for oil transport from the Middle East to China and Japan.

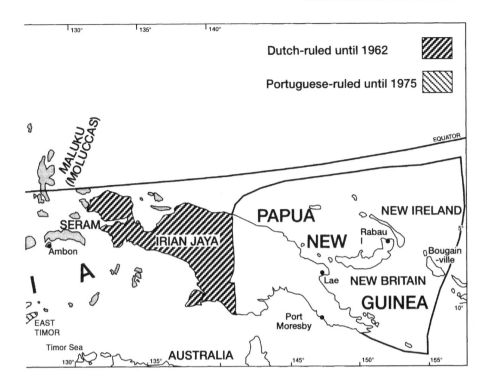

Papua New Guinea – the eastern half of New Guinea, with the adjacent islands, including New Britain, New Ireland and Bougainville – became independent in 1975. Its southern part had been, from 1884 to 1905, British New Guinea, and thereafter Australia's Territory of Papua. The northern part, under German rule as Kaiser Wilhelmsland from 1884 to 1914, later became an Australian-administered UN trust territory (which was occupied by Japanese forces from 1942 to 1944).

The 6 million people of Papua New Guinea (PNG) speak more than 700 different languages. As in Irian Jaya, the Melanesian population includes some inland tribes who, until recently, were completely isolated from the outside world. In the 1980s thousands of Melanesian refugees from Irian Jaya fled into PNG. Indonesian forces sometimes crossed the frontier in pursuit of rebel guerrillas.

Papua New Guinea has faced a separatist rebellion of its own, in Bougainville, which is geographically one of the Solomon Islands, but had become separated from the rest of the group by a boundary line originally drawn in the 1880s between British and German possessions. In 1976 the island was granted a degree of autonomy, but a resurgence of unrest led to open rebellion in 1988. The Bougainville rebels' activity stopped work at a copper mine which had been providing a third of PNG's export earnings. A ceasefire in 1997 was followed by a peace agreement in 2001 promising a future referendum on independence.

65 Australia and New Zealand

These two nations, separated by 1,000 miles of the Tasman Sea, are isolated but also shielded by wide oceans on three sides. Mainly peopled from Britain, they at first relied on British naval protection; in 1914 and again in 1939 they rallied to Britain's side in war, sending forces to fight in Europe and the Middle East. But in 1942 they themselves faced a threat from the north: Japan's advancing forces were not stopped until they had reached New Guinea and the Solomon Islands.

Australia and New Zealand made a defence treaty with the United States (ANZUS) in 1951; they signed the SEATO treaty in 1954, and they sent forces to Korea in the 1950s and to Vietnam in the 1960s (59, 61). But in 1984 New Zealand adopted an 'anti-nuclear' policy which led it to exclude US warships from its ports; this ended its participation in ANZUS. In 1985, French secret agents sank, in Auckland harbour, a ship which an anti-nuclear group was planning to take to French Polynesia in a bid to impede test explosions there (66). A period of strain in French–New Zealand relations followed.

Most of the 20 million Australians are townspeople. Half of them live in the five biggest cities, which are all on the coast of this huge country (or continent); a large part of its interior is desert or arid near-desert. Its formerly pastoral economy has been transformed, largely by discoveries of great mineral wealth; minerals, including coal, now provide half of its exports. Minerals not marked on the map include copper, gold, tin and tungsten.

Two-thirds of New Zealand's 4 million people now live in the main urban areas, but its economy is still largely pastoral, wool, meat and dairy products forming about 35%

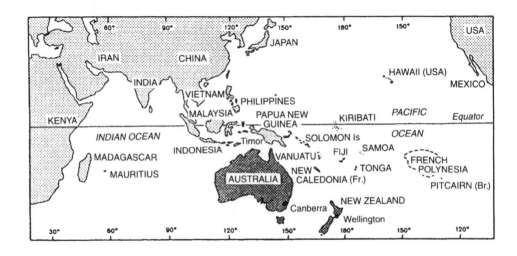

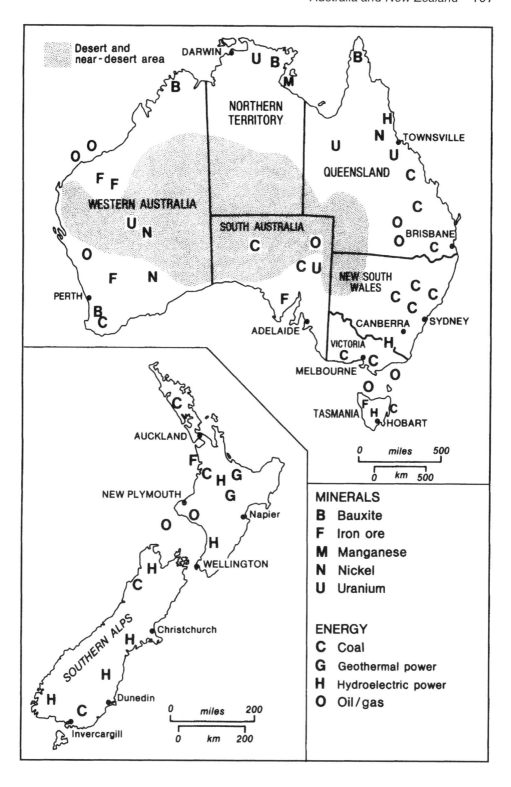

Desert and near-desert area

DARWIN

U B
B
M

NORTHERN
TERRITORY

O O
F F

WESTERN AUSTRALIA

U N

O
F N

PERTH
B
C

H
N
TOWNSVILLE

U

QUEENSLAND

C

C

O
O
BRISBANE
C

SOUTH AUSTRALIA
C

O
C U

NEW SOUTH
WALES
C
C C
C C
CANBERRA
SYDNEY

F

ADELAIDE

VICTORIA
H

C
C
O

MELBOURNE
O

TASMANIA
F
H
C
HOBART

| 0 | miles | 500 |
| 0 | km | 500 |

AUCKLAND
C

F
C H G
NEW PLYMOUTH
G

O O
Napier

H

C

WELLINGTON

SOUTHERN ALPS

H
Christchurch

H

H
Dunedin
C
Invercargill

| 0 | miles | 200 |
| 0 | km | 200 |

MINERALS
B Bauxite
F Iron ore
M Manganese
N Nickel
U Uranium

ENERGY
C Coal
G Geothermal power
H Hydroelectric power
O Oil/gas

of its exports. In 1973 it faced a struggle to diversify its trade quickly when Britain, which used to take 70% of its exports, joined the European Community, whose protectionist farm policies had the effect of slashing New Zealand's sales of meat and butter to Britain. Its biggest markets are now Australia, Japan and the United States; only about a sixth of its exports go to EU member countries.

In their 1965 free trade agreement Australia and New Zealand made a start on reducing tariffs on trade between them. In 1982 they completed their 'closer economic relations' (CER) pact, with the aim of progressively removing nearly all restrictions on this trade. Meanwhile, reflecting the changing focus of its trade, foreign policy and immigration, there has been a shift towards identification of Australia as a stable Asian country rather than an outpost of Europe in the Pacific. Australian gold, coal, wool and other commodities have found ready markets in Asia, which takes half of Australia's exports. Asians, including many refugees from Vietnam *(60)*, now make up 7% of its population and 40% of its annual intake of immigrants (the old 'white Australia' policy was abandoned in 1973). Australia has long been the 'superpower' of the Pacific Islands. More recently, with instability and terrorist attacks in Indonesia, Australia has expanded its role in Asia. In 1999 it led the peacekeeping force that preceded the independence of East Timor *(64)*, and it sent troops to Afghanistan and Iraq following the recent American invasions *(48, 50)*.

Only about 500,000 Australians are of 'Aboriginal' descent; but 15% of New Zealanders are at least partly of Maori (Polynesian) origin. There has also been much immigration into New Zealand of other Polynesian islanders, especially from the Cook Islands and Niue, former dependencies now in association with New Zealand, and Samoa, a former UN trust territory now independent *(66)*. In Auckland, the biggest city, a third of the inhabitants are Polynesian or Asian. In 1975 a tribunal was formed to resolve Maori claims for return of ancestral land or compensation under the terms of the 1840 Treaty of Waitangi.

66 South Pacific

Tiny Pitcairn is now this region's one remaining British dependency. Between 1970 and 1980 independence was achieved by Fiji, Tonga and the Solomon Islands, and by Kiribati and Tuvalu (formerly linked as the Gilbert and Ellice Islands). Banaba (Ocean Island) was included in Kiribati after long negotiations with the 2,000 Banabans, who had sought separate independence.

French Polynesia, New Caledonia, and Wallis and Futuna are still French overseas territories. In New Caledonia, unrest has sometimes taken violent forms among the Kanaks (Melanesians), who now form only 43% of its population (37% are Europeans, many of them connected with the island's role as a major producer of nickel). In the 1980s changes of government in France produced several changes in the constitutional arrangements proposed for New Caledonia. In French Polynesia there were protests about France's use of Mururoa atoll for nuclear tests *(4)*; but in 1996 France promised to hold no more tests, and signed the relevant protocols to the 1985 Rarotonga treaty.

In 1980 the New Hebrides, a British–French condominium, became independent Vanuatu. It then faced a secessionist rebellion on Espiritu Santo island, and it called in troops from Papua New Guinea to help suppress it. The Cook Islands and Niue are self-governing in association with New Zealand; the Tokelaus are a New Zealand territory.

Nauru and (formerly Western) Samoa, previously UN trust territories administered by Australia and New Zealand, are also now independent. Nauru, like Banaba, was devastated by twentieth-century phosphate mining. In 1993, after pursuing damage compensation and additional royalties in the International Court of Justice, Nauru gained a settlement from Australia, New Zealand and Britain. As part of a 'Pacific Strategy' after the 9/11 terrorist attacks, Australia opened offshore centres for processing its illegal immigrants in Nauru and Papua New Guinea.

Global warming poses a particular threat to island nations whose entire territory is threatened by rising sea levels. It could force abandonment of countries such as Tuvalu, whose maximum elevation is 16 feet above sea level.

The region's most populous state is *Fiji*, with 900,000 people, many of them descended from immigrants who came from India a century ago. When the 1987 election brought an Indian-led government to office, the army seized power and a new constitution was imposed to ensure continuing Fijian dominance. It was this overt racial discrimination that led the Commonwealth to suspend Fiji from membership *(9)* until 1997 when the constitution was revised. The election victory of an Indo-Fijian led to another coup and Commonwealth suspension in 2000. This was followed by further elections in 2001 and Fiji's readmission to the Commonwealth. Yet another coup in 2006 caused Fiji's third suspension. So many Indians have now left Fiji that they no longer form more than half of the

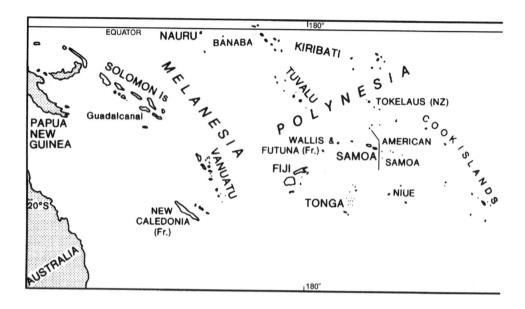

population. Despite its domestic difficulties, Fiji's contribution to UN peacekeeping operations has been out of all proportion to its size.

Since 1971 the island states and territories have met regularly with Australia and New
Zealand in the South Pacific Forum. In 1980 its 15 members signed a regional trade and
economic co-operation agreement (SPARTECA) that gives island products favoured access
to the Australian and New Zealand markets. In 1985 they concluded the Rarotonga treaty,
designating the South Pacific as a nuclear-free zone but leaving open the question of visits
by nuclear-armed or nuclear-powered ships, on which New Zealand was in dispute with the
United States *(65)*.

North of the equator, but often called 'South Pacific' islands, are the Carolines, Marianas
and Marshalls, which had been held by Germany before 1914 and were then held from 1919
onwards by Japan; in 1947 they became an American-administered 'strategic' UN trust
territory, also called Micronesia *(58, 67)*. In the 1970s the Northern Marianas (including
Saipan) chose to become a commonwealth associated with the United States, with much
the same status as Puerto Rico *(73)*. The Marshalls and the Federated States of Micronesia
(most of the Carolines) chose independence in 1986, and Palau did so in 1994, but without
severing all links with the United States. The Americans retained the right to use some
defence facilities on Micronesian islands. Earlier, US missiles had been test-fired to sea
areas near Kwajalein in the Marshalls, and in the 1950s Bikini and Enewetak, also in the
Marshalls, had been used for US nuclear testing; these islands remained completely or
partly uninhabitable. The United States continued to provide the islands with economic aid.

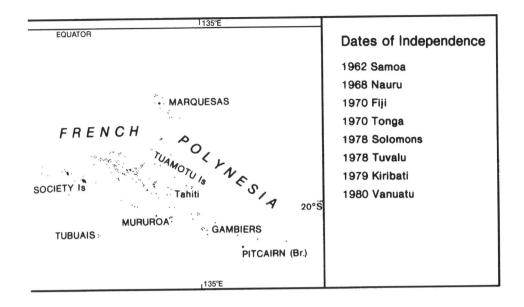

EQUATOR

135°E

MARQUESAS

FRENCH POLYNESIA

TUAMOTU Is

SOCIETY Is

Tahiti

20°S

MURUROA

GAMBIERS

TUBUAIS

PITCAIRN (Br.)

135°E

Dates of Independence

1962 Samoa
1968 Nauru
1970 Fiji
1970 Tonga
1978 Solomons
1978 Tuvalu
1979 Kiribati
1980 Vanuatu

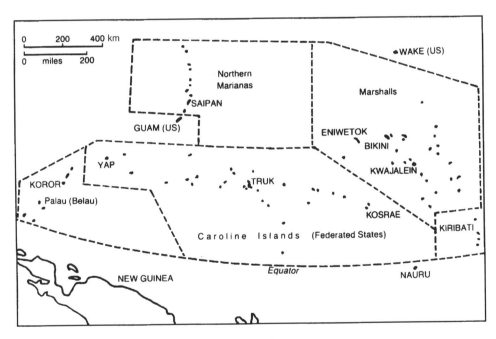

0   200   400 km

0   miles   200

WAKE (US)

Northern
Marianas

Marshalls

SAIPAN

GUAM (US)

ENIWETOK

BIKINI

YAP

KWAJALEIN

KOROR

TRUK

Palau (Belau)

KOSRAE

KIRIBATI

Caroline Islands (Federated States)

Equator

NEW GUINEA

NAURU

67 America and the Pacific

In the nineteenth century the Americans reached out into the Pacific. They bought Alaska from the Russians (1867), took the Philippines and Guam from Spain (1898), and annexed Hawaii, Midway, Wake and eastern Samoa. Between 1904 and 1914 they built the Panama Canal, giving ships a far shorter route between the Atlantic and the Pacific than the ones through the Magellan strait or round Cape Horn *(5, 72)*. In Japan, whose rulers had long

rebuffed all foreign contacts, the Americans' opening up of the country to trade in 1854 had fateful consequences; later they became alarmed by Japan's ambitions, which, in due time, led to its 1941 attack on the US naval base at Pearl Harbor, on Oahu island in the Hawaiian group. Japan then seized the Philippines, Guam, Wake and the westernmost Aleutians *(58)*.

After the Americans had fought their way back across the Pacific and forced Japan to surrender in 1945, they became committed to defending areas – including Japan itself – on the western side of the ocean. In the 1950s the invasion of South Korea and the communist forces' advances in Indochina aroused fears about the new power of a China which then had Soviet backing. The Americans undertook to protect Taiwan against attack from the Chinese mainland, and South Korea against any renewed attack from the north; and they were gradually drawn into the conflict in Vietnam *(56, 59–61)*.

In 1951 they joined Australia and New Zealand in the ANZUS treaty and signed a defence treaty with Japan *(58, 65)*. In 1954 the SEATO treaty was signed at Manila by the ANZUS allies and Britain, France, Pakistan, the Philippines and Thailand *(60)*. But the whole situation in the western Pacific region was changed in the 1960s and 1970s by the rift between China and the USSR, the American rapprochement with China, the communists' military victories in Indochina, and the bitter feelings that the war in Vietnam aroused

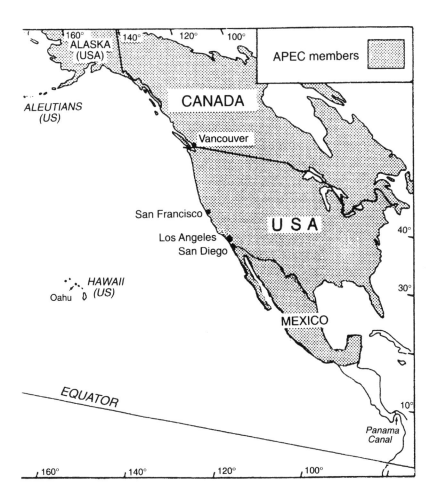

among Americans, moving them to withdraw from that struggle and to recoil from the idea of involvement in any similar lengthy conflict on the Asian mainland *(54, 61)*. SEATO was terminated in 1977; and the American bases in Thailand were relinquished (but those in the Philippines were retained until 1992).

By the mid-1990s there had been further dramatic changes. The breakup of the USSR had led to the withdrawal of Soviet warships from the Pacific, and the withdrawal of Soviet support from North Korea. Japan had become an economic superpower – although American forces remained both there and in South Korea, where there were new fears about North Korea's bellicosity and its attempts to build nuclear weapons *(59)*. Among the members of the Asia–Pacific Economic Co-operation forum (APEC), created in 1989 *(2)*, China and Japan were prominent, alongside the United States – with which, at times, each of them had sharp disputes about trade.

The Americans had given the Philippines independence in 1946. Between 1953 and 1972 they returned the Ryukyu, Bonin and Volcano islands to Japanese control *(58)*. Alaska and Hawaii are American states, and Guam, the Northern Marianas, American Samoa, Midway and Wake are US territories.

68 United States of America

There were originally only 13 states (all along the eastern seaboard) in the American federation. After 1912, when Arizona attained statehood, there were 48. In 1959, Alaska and Hawaii, formerly US territories, became the forty-ninth and fiftieth states.

Between the 1950 and 2000 censuses the population increased by 130 million; it is now over 300 million. The rapid increase has largely been the result of continuing immigration (some of it illegal, particularly across the Mexican border). Until 1965 immigration quotas were designed to preserve a predominantly European population mix, but in recent years about 80% of immigrants have come from the 'third world' (6). At 13%, the current share of foreign-born population is close to the highest level attained during the peak of European immigration a century ago.

The total population now includes about 45 million people of Hispanic American origin clustered in the south-west, in Florida and in the large northern cities; 14 million Asians, of whom half live in California, Hawaii or New York; 3 million American Indians, some of whom live on reservations, the remaining Indian-controlled territory comprising 3% of the US land area; and 40 million African Americans, partly or wholly of sub-Saharan African origin. In the 1920s, 85% of black Americans were living in the south-east (the 'Old South'), but only half of them live there now, millions having moved to northern and western cities. In some cities and suburban areas there are black majorities, notably in the capital, Washington, and in Atlanta and Detroit.

The biggest shift of balance has been away from the north central ('Midwest') and north-east regions. Since the 1970s the south and west have contained more than half of the total population; they now elect more than half of both Houses of Congress, and have more than half of the votes in presidential elections. This shift represents economic as well as political change. Sophisticated industries have expanded in the far west (e.g. computers in 'Silicon Valley' south-east of San Francisco, jumbo jets in Seattle) and in the southern 'Sunbelt'. Traditionally it has been the young who migrate, but population growth in these regions has been strongly boosted by movement of retired people – the three fastest-growing states (Nevada, Arizona, and Florida) all attract elderly migrants. In the south, income per head was only 50% of the national average in 1940; now it is over 90%. New York, which until the 1960s was the most populous state, has been far outstripped by California, which had 11 million inhabitants in 1950 but now has 37 million; the Hispanic and Asian contributions to its population are more than twice as large as the national averages.

Population shifts have brought more Americans into the path of natural disasters – among them hurricanes in the south, earthquakes in the west, and tornados in the south and midwest. In 2005 a hurricane flooded and destroyed much of New Orleans, forcing almost

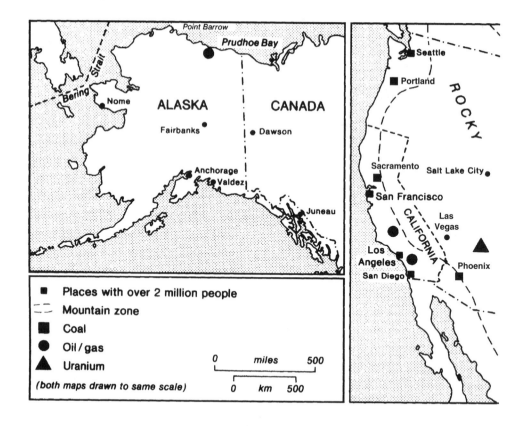

- ■ **Places with over 2 million people**
- - - **Mountain zone**
- ■ **Coal**
- ● **Oil / gas**
- ▲ **Uranium**

(both maps drawn to same scale)

complete evacuation; rebuilding continued more than a year later, but many people had left permanently.

In the south and west, development has been fostered by the exploitation of oil and natural gas and the building of dams (to provide both hydroelectric power and irrigation). The United States is a major oil importer, but Alaskan oil, piped to Valdez and thence moved in tankers, has helped to reduce its dependence on oil from the Middle East *(3, 41)*. In the 1980s the increase in its use of nuclear energy was sharply checked by fears about safety, partly caused by the 1979 accident at the Three Mile Island reactor in Pennsylvania, 100 miles west of Philadelphia.

For a time, while the west and the south were attaining their new prosperity, the older northern manufacturing areas, whose coalfields had long served a massive steel industry, became labelled as 'the Rust Belt'. But there was an impressive revival in the late 1980s and early 1990s in the 'Midwest' (roughly, the region between Minneapolis and Cleveland). By 1996 its exports were booming, and its unemployment rate remained low through 2005. The rate was low for several years.

Long home to the world's largest economy, the US dominates trade with its neighbours. The relationship between the US and Canada is the world's largest; two-nation merchandise trade equals that between the entire EU and the US. Trade with Mexico is half as large. In 1989 a free trade agreement with Canada came into effect. Tariffs on Canada–US trade were to be abolished within ten years. This agreement was merged into the North American Free Trade Agreement (NAFTA), which brought Mexico in as third party in 1994.

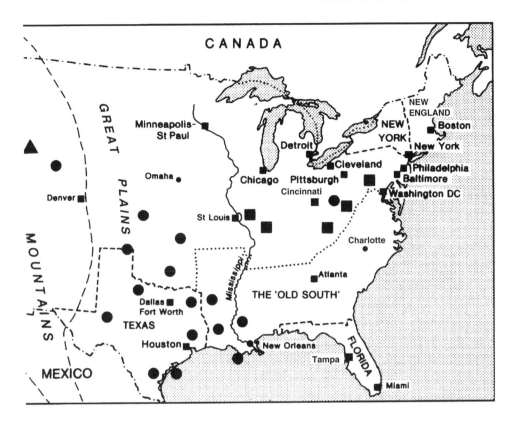

After the 1914–18 war, 'isolationism' had been strong enough to keep the United States out of the League of Nations. After the 1939–45 war, however, the headquarters of the League's successor, the United Nations, was sited in New York. As the greatest economic power, and the only military power able to match the Soviet Union, America found many other countries looking to it for help and leadership (although not necessarily following every lead it gave).

For fifty years it has kept forces and supported allies in Europe and the Far East *(11, 67)*; it has played a major part in Middle Eastern affairs *(42–9)*; more recently, it has taken a hand in tackling problems in several parts of Africa. Its international involvement, from 1945 onwards, had largely been a response to challenges from communist regimes in Europe and Asia. But neither in the late 1970s, after the bitter experience of the Vietnam war, nor in the early 1990s, after the collapse of communism in eastern Europe and the breakup of the Soviet Union, was there a reversion to 'isolationism' of the kind seen in the 1920s and the 1930s. Indeed, the 'war on terror' that began following attacks on New York and Washington in 2001 *(8)* has increased American involvement in the Middle East and else-where, while the status of the USA as the single global superpower is ensured by its half-share of the world's annual military budget.

69 Canada

Canada's existence has long been a defiance of both geography and history. Nearly all the 30 million Canadians live in a 3,000-mile-long strip of land bordering the United States; but, despite strong American influences, they have maintained a separate identity. In a sense, however, it has been a double identity. Historically, Canada's origins were bilingual. About 30% of Canadians speak French; around 40% are of British or Irish origin, and most immigrants of other origins have chosen to learn and speak English. 'Visible minorities' of primarily Asian origin have increased through immigration to 20% of the population.

Around the middle of the twentieth century, French-Canadians became more and more disturbed by the prospect of their distinctive culture being swamped in a predominantly English-speaking North America. Some of them campaigned to defend and strengthen their community's position throughout Canada; as a result, the country's federal governments took a series of actions to promote bilingualism and biculturalism. Although this had considerable effect, it was regarded as a lost cause by many people in Quebec, the only province with a French-speaking majority (three-quarters of all French-Canadians live in Quebec). They argued that French culture must be preserved in its one stronghold – if necessary by separating Quebec from Canada.

By the mid-1990s it seemed that Canada's unity was in doubt. A provincial referendum held in Quebec in 1995 narrowly failed to produce a mandate for secession; but, while the total voting was almost equally divided, among the French-speakers 60% voted 'yes'. Support for secession had clearly increased since the 1980 referendum, when it was the French-speakers who were almost equally divided, and the English-speakers' votes tipped the balance, producing a 'no' majority. And by now many Quebec English-speakers, finding that new language laws imposed by its provincial government left them at a disadvantage, were moving to other provinces.

Quebec separatism had been encouraged in the 1960s and 1970s by gestures from France, which strained relations between the governments in Ottawa and Paris. Provincial elections gave Quebec a separatist-minded government in the period 1976–85 and again from 1994 to 2003. Meanwhile a series of attempts were being made by federal governments to reduce restiveness in Quebec by constitutional changes and other moves.

In 1982, Canada at last completed its federal constitution – which dated from 1867 – by adopting a formula for amending it. (Until then, any change had to be formally approved by Britain, because this independent nation's federal government and its ten provincial ones could not agree on an amending process.) But Quebec spurned the 1982 agreement. In 1987, at talks held at Meech Lake, north of Ottawa, all provinces were offered wider powers; Quebec accepted the deal, but two provinces refused to ratify it. In 1992 a new

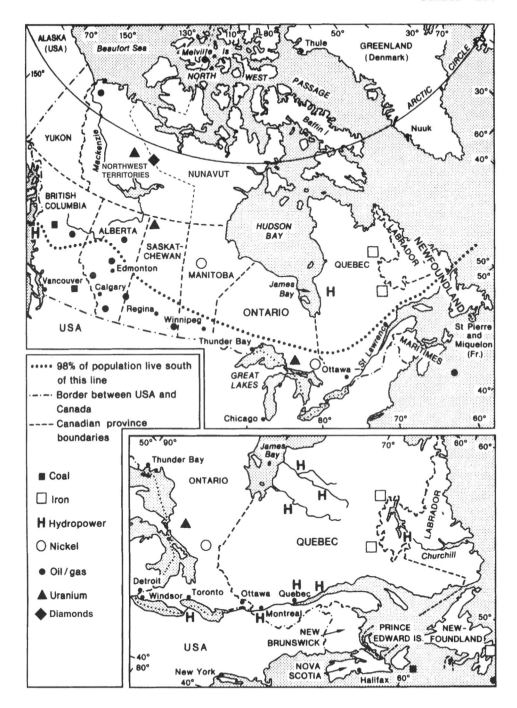

formula which had been agreed between all eleven governments (at Charlottetown, on Prince Edward Island) was rejected in a referendum by Quebec, Nova Scotia and the four western provinces. The federal government's attempts to satisfy Quebec's demands had produced an angry backlash, and even separatist rumblings, in the west. An oil-based economic boom in Alberta and migration from eastern Canada and Asia have increased the importance of the west relative to the traditional centres of eastern Canada.

Canada's aboriginal peoples (Indians and Inuit, numbering 1.3 million, or 4% of the population) also began to press their claims more vigorously. Meech Lake had been criticized for ignoring Indian and Inuit (Eskimo) rights. Plans were drafted for a division of the thinly peopled Northwest Territories into a western, Indian-oriented region (Denendeh) and an eastern, mainly Inuit one (Nunavut), each to be given increased powers of self-government. Nunavut was established in 1999 with a population of 25,000 in a territory of 700,000 square miles.

One of the four Atlantic provinces, Newfoundland, which had been a British dependency, joined the Canadian federation only in 1949. Off its coast lie the islands of St Pierre and Miquelon, an overseas *département* of France. Canada and France were long in dispute over fishing rights in these waters; an international tribunal provided a settlement in 1992. A similar dispute with the United States had been resolved in 1984. In 1995, Canadian and Spanish gunboats confronted each other when Canada took new action to curb overfishing off Newfoundland.

In 1989, Canada and the US joined in a free trade agreement, later expanded to include Mexico *(68)*. Canada and Mexico are the largest suppliers of oil to the US, and diamond mining has begun in Canada's Arctic territories.

70 Mexico

Mexico gained independence from Spain in 1810 but lost more than half its territory to Texan independence in 1836 (Texas joined the US in 1845) and war with the US a decade later. After the 1910 overthrow of dictatorship in the Mexican Revolution, Mexico became a one-party state under the Institutional Revolutionary Party. The foreign debt crisis of the 1980s *(74)* and recession following devaluation of the peso in 1994 gave way to rapid economic growth after the North American Free Trade Agreement (NAFTA) came into effect in 1994. Multiparty democracy arrived in the late 1990s.

In population, land area and GDP, Mexico ranks among the top 15 countries in the world. Of the 110 million Mexicans, a quarter live in the capital and surrounding states. More than 10% of Mexicans are members of indigenous groups, clustered in the south where many people speak Maya languages. From the brutality of the Spanish conquest of the Aztec and Maya regions to land disputes of the twentieth century, Indians have

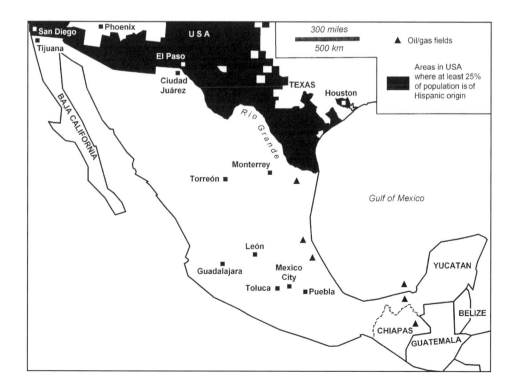

suffered from outside invasion and intervention. This is reflected in higher levels of poverty and illiteracy in Maya-populated areas today. In 1994 a guerrilla group, the Zapatista National Liberation Front (named after a leader of the 1910 revolution), demanded greater autonomy for indigenous groups. The Mexican government first attempted to suppress the rebellion through military means, then began negotiation in 1995.

In joining the North American Free Trade Agreement in 1994, Mexico tied its economy even more closely than before to that of the US. Investment in factories assembling goods for export to the US (*maquiladoras*) has brought jobs to the poor US–Mexico border region. Economic diversification means that petroleum now accounts for a much smaller share of Mexican exports than in the 1970s, when discoveries in the south first made Mexico a large oil exporter. Rapid population increase on both sides of the border strains water supplies in an arid region, and many factories have been criticized for their working conditions or lack of pollution controls.

Mexico has been the largest source of immigration to the US since the 1960s. The Mexican-origin population in the US is equivalent to a quarter of Mexico's own population. Most Mexican-Americans live in the south-western states of the US that were part of Mexico until the mid-nineteenth century. Neither NAFTA and economic growth, nor increased fortification of parts of the border, nor slower population growth since 1980 have deterred migration. Hundreds of thousands of Mexicans (and Central Americans) cross the US–Mexico border every year. Except for a small flow of retirees, very few American citizens seek to migrate the other way.

71 Central America, Caribbean, Cuba

The United States got Panama to break away from Colombia in 1903, and then built the Panama Canal *(72)*. Other small ex-Spanish states in the region also became, in effect, protectorates of the US, which, until the 1930s, repeatedly sent troops to stop civil wars or restore order. Several of these states became notorious for oppressive military or right-wing rule (Costa Rica, which has no army, is a notable exception). In 1954 a leftist government in Guatemala was ousted by rightist exiles who, with American backing, launched an invasion from Honduras. In 1965, US forces were sent to stop a civil war in the Dominican Republic, but were soon replaced by troops from other OAS countries *(74)*, who also withdrew after elections had been held.

In *Haiti* the Duvalier family's 29 years of power ended in 1986, but the military and police chiefs prevented the creation of a democratic government; in 1991 they ousted a newly elected president. Their hold was broken in 1994 by the arrival of US troops, replaced six months later by a smaller United Nations force (including an American contingent). A civilian government was installed, but it was overthrown in 2004, and the UN returned. Haiti's woes, like Cuba's (see below), have led thousands of people to put out to sea, heading for the United States. Near-total deforestation and occasional hurricanes have increased soil erosion.

After failure of the rebellion of 1868, *Cuba* again sought independence from Spain in 1895 under José Martí. The US declared war on Spain in 1898. After the defeat of Spain, Cuba gained independence under American 'protection' and the US acquired a base at Guantánamo on a permanent lease. The US also took control of Puerto Rico, Guam and the Philippines. Dictatorial rulers, US military intervention, and American domination of the Cuban economy (especially sugar production) led to the civil war of 1956–9. This ended in victory for the forces of Fidel Castro, who nationalized US-owned businesses and created a communist regime. In 1961 it repelled a US-backed landing by Cuban anti-communists at the Bay of Pigs. In October 1962 the Americans detected Soviet preparations to site nuclear missiles in Cuba; after a confrontation that aroused widespread fears of a full-scale nuclear war, the Soviet ships carrying the missiles were turned back.

With Soviet aid, Castro built up a large army, and some of his troops were sent to Africa to fight 'proxy wars' for the USSR (in Angola and Ethiopia *[33, 35]*) between 1975 and 1991. At home, while Cuba remained among the world's poorer nations, Castro raised literacy and life expectancy to first-world levels through emphasis on basic health care and education. After 1991 the loss of Soviet subsidies, combined with the US ban on trade and the communist economy's inefficiency, brought dire poverty to Cuba. Many people there depended on remittances sent by the 2 million Cubans who had escaped. The continuing flow of escapers – many of them desperate enough to try to float to Florida on rafts

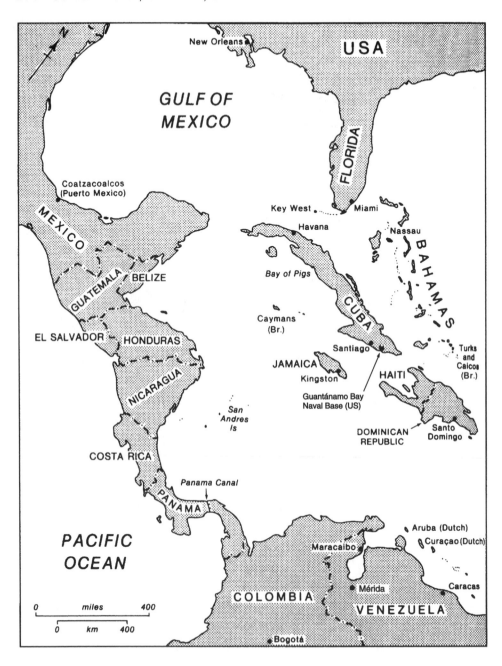

New Orleans
USA
GULF OF
MEXICO
FLORIDA
Coatzacoalcos
(Puerto Mexico)
Key West
Miami
MEXICO
Havana
Nassau
BAHAMAS
BELIZE
Bay of Pigs
GUATEMALA
Caymans
(Br.)
CUBA
EL SALVADOR
HONDURAS
Santiago
Turks
and
Caicos
(Br.)
JAMAICA
NICARAGUA
Kingston
HAITI
Guantánamo Bay
Naval Base (US)
San
Andres
Is
DOMINICAN
REPUBLIC
Santo
Domingo
COSTA RICA
Panama Canal
PANAMA
Aruba (Dutch)
Curaçao (Dutch)
PACIFIC
OCEAN
Maracaibo
Caracas
COLOMBIA
Mérida
0   miles   400
VENEZUELA
0   km   400
Bogotá

– caused recurrent problems for the United States. For a time, Guantánamo was used as a temporary holding station for thousands of the Cubans found out at sea in frail craft. After the 9/11 attacks *(8)*, it became an offshore prison for suspected terrorists.

The impact on the region of a Soviet-backed Cuba was not the only cause of tensions after 1959. A frontier dispute between Honduras and El Salvador led to a short war in 1969. A chain reaction of conflicts in Central America was set off in 1979, when the Somoza

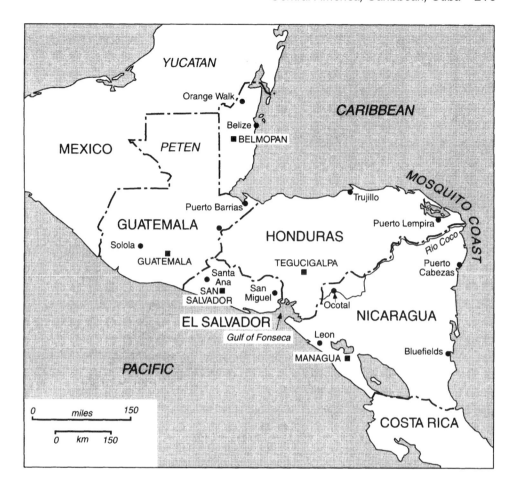

family, who had ruled *Nicaragua* dictatorially since 1936, were ousted by the Sandinistas, leftist guerrillas named after a rebel leader killed in 1934. Once in power, the Sandinistas broke with their non-communist supporters, imprisoned many of them, and sought Cuban and Soviet help. Their victory (and their covert aid) encouraged leftist guerrillas in *El Salvador* to begin a full-scale civil war in 1980. The army, with US support, employed 'death squads' that killed 30,000 people in 1979–81. There were attacks on Nicaragua from Honduras and Costa Rica by exiles, dubbed '*contras*' (counter-revolutionaries). In Guatemala another guerrilla struggle, marked by the army's brutality towards restive Maya Indian peasants, drove many refugees into Mexico.

In the early 1980s the United States became more and more openly involved in supporting the Nicaraguan '*contras*' and putting other pressure on the Sandinistas. At the same time it pressed the neighbouring states to end military rule: Honduras did so in 1981, El Salvador in 1984, Guatemala in 1985 (Belize and Costa Rica already had civilian governments). By 1987, American support for the '*contras*' had become restricted – and the covert operations intended to maintain them had been publicly exposed. The 'Contadora group' *(74)*, Costa Rica and others helped to frame a regional peace plan. Talks began between El Salvador's government and its guerrilla leaders; in 1992 a UN force was sent in and the twelve-year civil war, which had cost 75,000 lives, was ended.

In Nicaragua the internationally supervised arrangements brought, in 1990, the disarming and disbanding of the '*contras*' and the holding of a free election. The Sandinista junta's intolerant rule had lost it so much support that (to its surprise) its opponents won the election. The Sandinistas allowed an elected government to take over; but they retained command of the army through the mid-1990s.

In *Guatemala*, the US supported a coup in 1954 that removed a government bent on nationalizing the huge landholdings of American corporations. The region's longest civil war ensued, causing around 200,000 deaths from 1960, with the Maya Indians (half of the population) suffering especially from massacres and destruction of their villages. In 1996, after five years of negotiations, the rightist government and four leftist rebel groups signed a ceasefire and an agreement on constitutional reform, but little progress has been made since then.

Guatemala had long pressed claims to *Belize* (formerly British Honduras), whose independence was delayed until 1981 by Guatemalan threats to invade it as soon as the British left; after 1981 a small British force remained there at Belize's request. In 1991, however, Guatemala accepted Belize's independence and established diplomatic relations with it.

72 Colombia and Panama

The isthmus of Panama was part of Colombia in 1902. In 1903 the United States proposed the building of a canal across Panama, which would vastly shorten the sea route between the Atlantic and the Pacific. When Colombia was slow to agree, the Americans fomented a secession in Panama and prevented Colombia from suppressing it. Under a treaty with the new Panamanian republic, they built and operated the Panama Canal *(5, 67)* and controlled a 10-mile-wide zone along its banks. New treaties that took effect in 1979 abolished the Canal Zone and gave Panama more say in the operation of the canal. Panama took full control in 1999.

In 1989, American forces invaded Panama, where General Manuel Noriega (previously a US ally) had seized power in 1984, manipulating a series of figurehead presidents and rigging elections or simply overruling their results. An elected civilian government was installed, and Noriega, who had been indicted in the American courts on charges of helping Colombian 'drug barons' to smuggle narcotics into the United States, was taken there, tried and imprisoned.

During the 1980s there was a huge increase in Colombian production of cocaine, from coca grown locally and in Peru and Bolivia, and in smuggling to North American and European markets. By the end of the 1980s large amounts of heroin were also being produced from opium poppies grown in Colombia, Peru and Bolivia. Drug barons based in Medellín and Cali had made deals with leftist guerrilla gangs, and were bribing, terrorizing or killing policemen, officials, judges and politicians, frustrating attempts to arrest drug traffickers and to extradite some of them from Colombia to the United States.

The presidents of Colombia, Peru, Bolivia and the United States held an 'anti-cocaine summit' in 1990 at Cartagena and stepped up joint action, but the drug producers and smugglers defied their efforts. When the 'Medellín cartel' was broken in 1993, Cali became the traffickers' main base; it was reckoned that Cali handled three-quarters of all the cocaine entering the United States. More recently, the Mexican border has become the primary transit point for American cocaine imports, following action against smuggling in the Caribbean.

Since 2000 the US has funded 'Plan Colombia', a programme of military and economic assistance aimed at reducing coca production. Some reduction in coca-farming was achieved through airborne spraying of herbicide, at the cost of destroying other crops and exposing the rural population to chemical pollution. The Plan tackles only production, not demand; reduction in Colombian production has led to increased coca cultivation in Bolivia and Peru. Eradication of coca is complicated by traditional Andean use of the leaf (not refined cocaine) as a mild stimulant, and continued demand in North America and Europe ensures that coca cultivation is profitable for farmers.

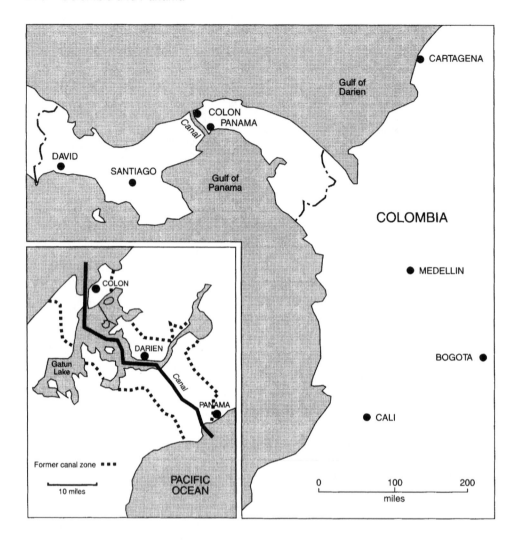

Left-wing guerrillas and right-wing paramilitaries have been active in Colombia for more than 40 years. The two main guerrilla groups, the FARC (Revolutionary Armed Forces of Colombia) and the ELN (National Liberation Army), oppose the government and foreign investment (especially in oil extraction), both of which are perceived to benefit only the wealthy. They are of Marxist origin and have received Cuban support. The ELN specializes in kidnapping for ransom, while the FARC receives much of its revenue from involvement in the drug trade, including 'taxes' on drug production and smuggling. It has used its profits from these activities to purchase modern weapons and has gained further supplies through attacks on army and police posts. Besides kidnap and murder, it has carried out terrorist bombings and has had ties with the Irish Republican Army *(24)*. The guerrillas are most active in the south and in the north near (and across) the Venezuelan border.

In 1997, because activities of the FARC cut into the drug trade, several paramilitary groups joined forces with leading drug traffickers in the AUC (United Self-Defence Forces of Colombia), which combats the guerrillas by massacring peasants and members of indigenous groups suspected of supporting the guerrillas. Indian communities have especially opposed government grants of oil-exploration licences in or near their territories and the resulting land invasions. The AUC has at times allied itself with the Colombian military, itself accused of serious human rights abuses.

In 1998 the Colombian government withdrew its army from a large area in the south to encourage talks with the FARC, but after continued conflict the government invaded the demilitarized area in 2002. Talks with the AUC led to disarming of some members starting in 2003, and in 2005 further negotiations began between the government and the FARC and the ELN.

73 East Caribbean, Guianas, Venezuela

Britain had been encouraging its Caribbean colonies to unite on their way to independence, but the West Indian Federation formed in 1958 broke up in 1962. By 1983, 12 British colonies had become independent as separate states *(9, 73)*; however, they co-operated in the Caribbean Community (Caricom), formed in 1973. In 1994 this group joined several mainland countries in a new Association of Caribbean States.

In 1981 the Organization of Eastern Caribbean States (OECS) was formed by Antigua and Barbuda, Dominica, Grenada, Montserrat, St Kitts and Nevis, St Lucia, and St Vincent and the Grenadines. These ex-British West Indian islands are now independent – except Montserrat, whose 10,000 inhabitants live adjacent to an active volcano – as are also Trinidad (and Tobago), Barbados and Guyana *(9)*. Anguilla, formerly linked with St Kitts, broke away and chose in 1971 to remain a British dependency.

The small sovereign states of the east Caribbean are intermingled with American, British, Dutch and French islands. Guadeloupe (with St Barthélemy and part of St Martin) and Martinique are French overseas *départements*; so is French Guiana (Guyane) on the mainland. Suriname (ex-Dutch Guiana) became independent in 1975. All the Dutch islands, except Aruba, form the self-governing Netherlands Antilles, with its capital at Willemstad on Curaçao. Aruba broke away in 1986, aiming to become independent by 1996; but in 1990 it abandoned that aim.

In *Grenada*, a coup in 1979 installed a leftist regime which broke its promise to hold elections and obtained support from Cuba. In 1983 its ruling junta split, and it collapsed in bloody disarray. The OECS appealed for help, and troops from the United States and six Caribbean islands landed, meeting resistance mainly from armed Cubans. An election in 1984 restored democratic government.

The other ex-British islands held free elections regularly. In 1981 a coup was attempted in Dominica with the help of foreign mercenaries. In 1990 a bid for power was made in Trinidad by a group of self-styled 'Muslims' (unconnected with the island's many Muslims of Asian origin). Both attempts were quickly suppressed. Trinidad and Tobago is one of the largest exporters of natural gas to the US.

Venezuela claims the part of *Guyana* west of the Essequibo river, rejecting its 1899 award to Britain by a European and American arbitration panel. Guyana also has a border dispute with neighbouring *Suriname*, and Suriname with French Guiana. Guyana has been troubled by its racial divisions; the ancestors of about half of its people came from East Asia, those of the rest from Africa. Suriname has similar problems; half of its population is of (East) Indian or Javanese origin, and most of the other half have at least partly African ancestry. In 1980 its small army seized power, and ruled so brutally that a third of the population fled

to Holland; civilian government was restored in 1988, overthrown by a second military coup in 1990, but restored again in 1991. Satellites are launched from French Guiana by the European Space Agency.

Two military coups were attempted in 1992 in Venezuela, the largest oil exporter in the Americas. Their leader, Hugo Chávez, gained the support of the poor and was elected president in 1998. He supplied Cuba with oil, expanded basic health care and education, and proclaimed his support for Fidel Castro. Chávez survived a coup attempt in 2002 and a referendum in 2004, and – while maintaining oil exports to the United States – sought allies among left-wing Latin American leaders. He opposed the Free Trade Area of the Americas *(74)* and sought to export his 'Bolivarian revolution', named after the Venezuelan Simón Bolívar, who led several nations to independence from Spain in the nineteenth century and was an early opponent of slavery.

74 Latin America

Brazil represents what was Portugal's empire in the Americas. The former Spanish empire is represented by the 18 Hispanic (Spanish-speaking) republics; they include two in the Caribbean: Cuba and the Dominican Republic. Haiti was once a French possession. These 20 Latin republics, with the United States, were members of the Pan-American Union, which was succeeded in 1948 by the Organization of American States (OAS); in 1947 they concluded the Rio treaty, providing for joint action if any of the 21 members was attacked. In 1962, Cuba's participation in the OAS was suspended. Since 1991 all 35 independent American countries have been members.

Several economic groupings were formed in the 1960s, but with little effect. In the early 1990s real progress towards regional free trade began; the most significant moves were the creation of 'Mercosur' by Argentina, Brazil, Paraguay and Uruguay, and the inclusion of Mexico in the North American Free Trade Agreement *(2)*. A US–Central American Free Trade Agreement (CAFTA) had been ratified by most prospective members by 2005. A proposed Free Trade Area of the Americas (FTAA) – to include all countries bar Cuba – has been opposed by left-wing Latin American leaders who view the FTAA as an avenue for US domination of the region. Instead, talks began in 2004 to create a South American Community of Nations open to all of the continent's independent states. Most members of Mercosur became 'associate' members of the Andean Group *(2)* and vice versa; Chile also has ties to both. In 2006, Venezuela left the Andean Group to join Mercosur, and Bolivia sought to follow – leaving behind Colombia, Ecuador and Peru, all of which have free trade agreements with the US.

In 1984 the 'Cartagena group' was formed by 11 republics, for consultation about their problem of heavy foreign debts. This problem, intensified by over-eager borrowing, through banks, of the huge sums that some OPEC countries had piled up after the oil price rises of the 1970s *(3)*, was not limited to Latin America; but Brazil and Mexico alone owed about $200 billion, and it was Mexico that dramatized the crisis when it announced in 1982 that it could not keep up payments on its debts.

In the 'Contadora group', formed at a meeting on an island near Panama in 1983, Colombia, Mexico, Panama and Venezuela united to try to moderate the conflicts among the Central American states *(71)*. Later, Argentina, Brazil, Peru and Uruguay joined them in a 'Group of Eight', which also held meetings about economic problems.

One obstacle to regional unity has been the persistence of old territorial disputes. There has been no recent war comparable to the one between Bolivia and Paraguay from 1932 to 1935 (the 'Chaco War', named after the disputed area). But Argentina and Chile had a long quarrel over the 'Beagle Channel islands' *(75)*; and Venezuela made claims to part of Guyana *(73)*. Ever since the 1879–81 war in which Chile took from Peru and Bolivia the

USA

MEXICO

• Mexico City

Bermuda (Br.)

BELIZE
GUATEMALA
HONDURAS
EL SALVADOR
JAMAICA
NICARAGUA

CUBA

DOMINICAN
REPUBLIC

HAITI

COSTA RICA
Cartagena
PANAMA

BARBADOS
TRINIDAD AND TOBAGO

VENEZUELA
Bogotá
COLOMBIA

GUYANA
SURINAM
FRENCH GUIANA

ECUADOR
Quito

Amazon

EQUATOR

Lima •
PERU

BRAZIL

• La Paz
BOLIVIA
Arica •

• Brasilia

CHILE
ANDES
PARAGUAY

Asunción

São
Paulo •

Rio de
Janeiro •

Santiago •

ARGENTINA

Buenos
Aires •
URUGUAY
Montevideo

• Viedma

0   miles   1000

0   km   1000

Falklands

Cape Horn

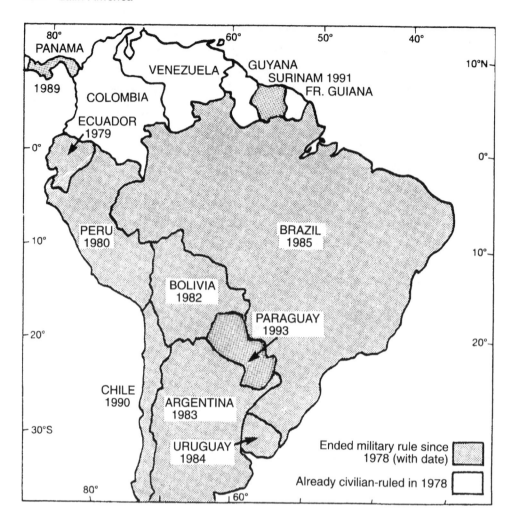

coastal area running south from Arica, Bolivia has sought to regain an outlet to the sea; talks in 1975 and 1987 failed to resolve this dispute, but in 1993 Peru offered Bolivia a free-trade port zone. In 1942, after a short war, Ecuador ceded territory to Peru; in 1961 it denounced the 1942 treaty; there were frontier clashes in 1981, 1984 and 1995, but a new treaty was signed in 1998.

The 1967 Tlatelolco treaty (named after a mountain near Mexico City) was designed to make Latin America a zone free of nuclear weapons. Argentina and Brazil, which then had secret plans to develop nuclear arsenals, held out, as did Chile; not until 1992 did the three agree to sign the treaty.

In the 1970s rule by generals who had seized power in military coups had been the predominant system of government in both South and Central America. During the 1980s there was an impressive swing to civilian government. Between 1979 and 1990, military rulers handed power over to elected civilian governments in Argentina, Bolivia, Brazil, Chile, Ecuador, Peru and Uruguay, and, among the Central American states, in Guatemala, Honduras, Nicaragua, Panama *(72)* and El Salvador.

Though the Latin American generals backed down, they did not lose all interest in political power. In *Paraguay*, General Alfredo Stroessner, who had ruled for 34 years, was ousted in 1989; another general took over, but promised to end military rule; an elected civilian president was inaugurated in 1993. Paraguay's generals permitted the 1993 election, but took care to influence its result. In Peru in 1992 an elected civilian president was able to dissolve the legislature and rule by decree because the army backed him.

In both *Peru* and *Bolivia*, the transition to democracy was accompanied by some improvement in the status of indigenous Andean Indians, guerrilla rebellions and American involvement. Peru elected its first Indian president in 2001, and Bolivia followed suit four years later. From 1980 to 1992, the Maoist Shining Path guerrilla movement was responsible for thousands of deaths in Peru, but most guerrillas surrendered after the capture of their leader in 1992. As in Colombia *(72)*, US-backed efforts to eliminate coca growing are complicated by the cultural and medicinal use of coca leaves in the Andes and continued foreign demand for cocaine. This was highlighted by the 2005 election of a former coca grower as president of Bolivia.

General Augusto Pinochet ruled *Chile* as a dictator from 1973 to 1990 after gaining power in a US-backed coup against an elected socialist president. He retained his command of the army until 1998. Pinochet's regime had been a brutally rightist one; a leftist mirror-image of his new position was created in Nicaragua when command of the army was retained by General Humberto Ortega, one of the leaders of its Sandinistas, after they had allowed an election to be held in 1990 and lost it. Both Pinochet and Ortega, on being urged to resign their commands, refused to budge. Since 1990, Chile has gained a reputation as a stable democracy. It signed a free trade agreement with the US in 2003 and has benefited from high prices for copper.

75 Argentina and Falklands

Independent for 190 years, Argentina, whose 40 million people are almost all of European origin (with big Italian and German elements), is not at all a typical 'Third-World' country. It has a large temperate zone, fertile plains, oil and other mineral resources, and good communications. Yet its economy is far less productive than that of, say, Canada – a country with a comparable population, similar resources and a harsher climate. The explanation

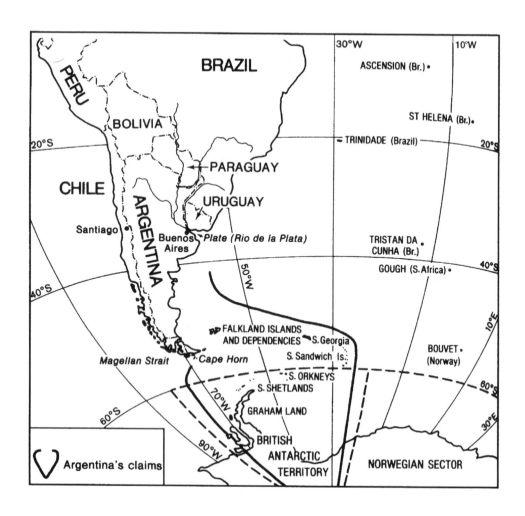

is, primarily, that Argentina has had a sorry record of misgovernment, both by some of its civilian governments and by all of its numerous military ones.

Geography has shielded Argentina from external threats; and in the 1914–18 and 1939–45 wars it remained lucratively neutral (with a tilt towards Germany in the 1939–45 war, after which it became a notorious refuge for Nazi war criminals). It has, however, maintained large armed forces, which have carried out many coups (e.g. in 1930, 1943, 1955, 1966 and 1976) and have influenced politics even when not openly holding power. And Argentine support has, at times, helped to sustain military regimes in neighbouring states such as Paraguay and Bolivia.

In the past half-century, Argentine ambitions have turned towards the south. After the exploitation of oilfields in Patagonia, hope of offshore oil has increased interest in the waters around Cape Horn, the Falklands and other southern islands. Like other signatories of the 1959 Antarctic Treaty *(76)*, Argentina has agreed, for the time being, not to press its claims beyond latitude 60° South, but there have been indications that it means to strengthen those claims with an eye to the future. Argentina and Chile, whose claims in Antarctica overlap, tend to use their armed forces to man their 'demilitarized' research bases there – and to fly officers' wives in to give birth there to future citizens who, being

Antarctic-born, may be eventually offered as 'living evidence' in support of territorial claims.

The introduction of 200-mile-wide sea zones *(5)* intensified an old quarrel between Argentina and Chile over three small islands – Lennox, Nueva and Picton, south of Tierra del Fuego – because possession of these islands could now determine the control of large adjacent sea areas. This was often called the 'Beagle Channel dispute', although the channel itself (named after the voyage in the 1830s of a British survey ship, *Beagle*, with Charles Darwin aboard) was not being contested; part of its course is a recognized frontier, and part is within Chile's territory.

In 1977 the three islands were awarded to Chile by five judges of the International Court, who had been asked to arbitrate in the dispute. Although Argentina had agreed to the arbitration, its ruling generals were so angry that for several months it seemed on the brink of war. The Pope was then asked for a decision; when, in 1980, he gave one in favour of Chile, Argentina again went back on its word and refused to give up its claims. Only after defeat in the Falklands had brought Argentina's generals down was a treaty, based on the Pope's proposals, concluded in 1984: Chile kept the islands, and a sea dividing line running south from Cape Horn was accepted.

The Falkland Islands are called Las Malvinas in Spanish; this is a borrowing from the French, who called them Les Malouines because their first pioneers there came from St-Malo in Brittany. The islands were uninhabited until settlements were founded by the French in 1764, the British in 1766 and the Spanish in 1767. The French renounced their claims and left; Britain withdrew its garrison in 1774 but maintained its claims; Spain abandoned the islands in 1810. Argentina made a claim to them in the 1820s but failed to take effective control. Britain reasserted its claims, effectively, in 1833, and a little British colony was soon thriving, servicing ships on voyages round the Horn. Port Stanley lost business to Punta Arenas in Chile when, as steam replaced sail, Chile's Magellan Strait replaced the Horn as the main route between the Atlantic and the Pacific; but in the years 1914–18 and 1939–45 the Falklands provided a valuable British refuelling and repairing port (the destruction of a German squadron there in 1914 was a turning-point in the naval war). Meanwhile a Falklands population of about 2,000, almost all of British origin, took to supporting itself mainly by sheep farming.

Argentina maintained its claims, but Britain promised the islanders that there would be no transfer of sovereignty without their consent. The two governments had for some time been holding talks about the future of the Falklands (with Britain trying to get an agreement on economic co-operation) when Argentina suddenly interrupted the talks by landing troops on the islands, and on South Georgia, in 1982. When attempts to negotiate a withdrawal failed, the British counterattacked. They quickly recaptured South Georgia and, landing in East Falkland, forced the invaders to surrender. The British task force had been operating at extreme range; the nearest airfield and staging-point available to it were at Ascension Island, 3,400 miles to the north.

Britain then found itself having to maintain an expensive garrison in the islands, and to build an airfield there fit for use by long-range airliners. In 1986, Britain proclaimed a 150-mile-wide fishery zone around the Falklands; this brought the islands new revenues from foreign fishing fleets, which had to buy permits.

After restoring diplomatic relations in 1990, Britain and Argentina agreed on joint action to control fishing in the waters between the mainland and the Falklands; an agreement about offshore oil was also negotiated. But there was an uncertain background to these developments. After defeat in the Falklands in 1982, Argentina's humiliated generals had to hand power to an elected government, which they then attempted to overthrow. The generals remained powerful enough to influence the civilian government, and also to shield from prosecution officers who had committed atrocities against fellow Argentines during the recent period of military rule.

Argentina's late-twentieth-century economic collapse would not have been predicted a century ago, when it was among the world's wealthiest countries. In 1992, after decades of economic mismanagement by dictators and episodes of high inflation, Argentina tied its currency to the US dollar and sought foreign investment. Growth returned for several years; but, when it slowed, the government found itself unable to pay its debts. The banking system fell apart in 2001, unemployment and emigration increased, and in 2002 Argentina defaulted on its World Bank debt.

76 Antarctic

Claims to sectors of Antarctica have been made by Australia, Britain, France, New Zealand and Norway, which recognize each other's claims, and by Argentina and Chile, whose claims are overlapping and also overlap Britain's. In 1959 these nations, and Belgium, Japan, South Africa, the Soviet Union and the United States, signed the Antarctic Treaty. Later, 33 other nations signed the treaty; half of these have mounted enough research activity to obtain full voting rights alongside the original twelve 'consultative parties'. The treaty banned military activity and new territorial claims. All research bases were to be open to inspection.

The treaty covered the area south of latitude 60° South. In 1980 the treaty states signed an agreement on conservation of living resources which covered the area encircled by the 'Antarctic Convergence' – the line where warmer water overlays the near-freezing Antarctic surface water. (These waters contain huge amounts of shrimp-like krill, mainly harvested by the Japanese and the Russians.) A 1991 protocol banned exploration for oil or minerals for at least 50 years.

Antarctica has no permanent population – only the teams manning the research bases maintained by America, Argentina, Australia, Brazil, Britain, Chile, China, France, Germany, India, Italy, Japan, South Korea, New Zealand, Poland, Russia, South Africa, Ukraine and Uruguay. These include an American base at the South Pole and a Russian base, Vostok, at the 'pole of inaccessibility'.

The hole in the earth's protective ozone layer caused by CFC (chlorofluorocarbon) emissions was discovered in 1982 by a British team at Halley. By 1995 teams at bases on the Antarctic Peninsula were reporting that rising temperatures had caused a crumbling of ice shelves that reduced their area by 3,300 square miles *(1)*. Melting of land-based ice in Antarctica and Greenland *(77)* – comprising the vast majority of the world's fresh water – would raise global sea level.

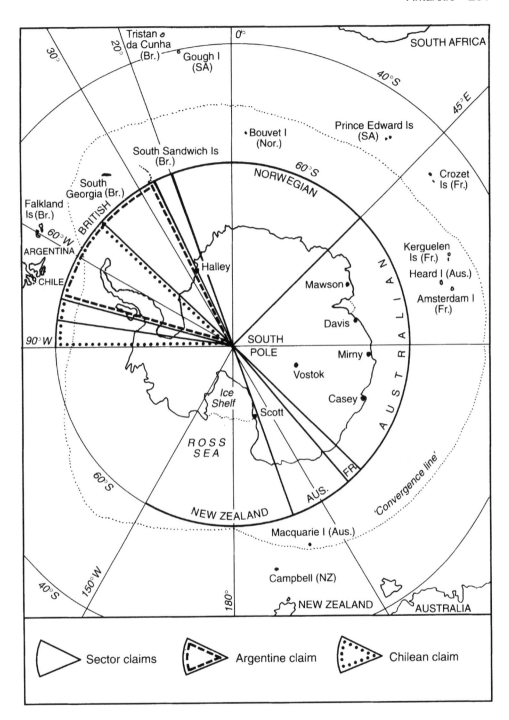

Tristan da Cunha (Br.)
0°
SOUTH AFRICA
Gough I (SA)
30°
20°
40°S
45°E
Prince Edward Is (SA)
Bouvet I (Nor.)
Crozet Is (Fr.)
South Sandwich Is (Br.)
60°S
NORWEGIAN
South Georgia (Br.)
Falkland Is (Br.)
BRITISH
Kerguelen Is (Fr.)
Heard I (Aus.)
60°W
Halley
ARGENTINA
CHILE
Mawson
Amsterdam I (Fr.)
AUSTRALIAN
Davis
SOUTH POLE
90°W
Mirny
Vostok
Ice Shelf
Casey
Scott
ROSS SEA
FR.
'Convergence line'
60°S
AUS.
NEW ZEALAND
Macquarie I (Aus.)
40°S
150°W
180°
Campbell (NZ)
NEW ZEALAND
AUSTRALIA

Sector claims
Argentine claim
Chilean claim

77 Arctic

In 1996 the Arctic Council was created by Canada, Denmark, Finland, Iceland, Norway, Sweden, Russia and the United States. One of its main concerns was environmental protection; a joint programme for that purpose had been initiated in 1991.

The first man at (or near) the North Pole was Robert Peary, who got there in 1909, using dog-drawn sledges. In 1958 other Americans could pass under the Pole in a nuclear-powered submarine which travelled beneath the ice from the Pacific to the Atlantic. By then, airliners were crossing the Arctic on flights between Europe, Japan and North America; and scientific posts had been set up on the ice of the polar region.

Coal has been mined in Svalbard by Norwegians and Russians. Oilfields have been exploited near Prudhoe Bay in Alaska, and in the Siberian region near the Ob river, from which gas pipelines to western Europe were built in the 1980s *(3)*. During the Soviet era, development in Arctic Siberia was largely based on the use of prisoners as slave labour, from the Vorkuta coalmines to the Kolyma goldfields, and the KGB ran whole districts as gigantic *gulags (16)*; Novaya Zemlya was a site for Soviet nuclear test explosions; and the bases around Murmansk were of great importance in Soviet naval strategy *(22)*.

Alaska, a territory which the United States bought from Russia in 1867, became the forty-ninth state of the US in 1959. Since mining began in the Northwest Territories in 1998, Canada has become one of the world's leading diamond producers. Iceland's constitutional links with Denmark ended when it became an independent republic in 1944. Greenland, a part of the Danish kingdom, was given 'home rule' status in 1979 and withdrew from the EU in 1985 *(12)*. Its population is only about 60,000 (Iceland has 300,000); its interior is covered by a huge ice cap, which rises to over 9,000 feet.

On the map the stippled areas show the difference between maximum and minimum (that is, midwinter and midsummer) ice coverage of the northern seas. In summer, ships equipped for ice-breaking can reach ports all along the Siberian coast; a 'north-west passage' through the Canadian Arctic islands has also been successfully navigated.

Within the Arctic Circle, until recently only one sea area was ice-free all year: the waters extending north from the Atlantic to Svalbard. Cold though these are, the Gulf Stream's continuation in the North Atlantic Drift brings relatively higher temperatures to them. Lately there has been an increase in summer ice-melting off Siberia and Canada – a symptom of global warming *(1)*. The disappearance of floating Arctic ice would not raise sea levels; but the shrinking of Greenland's ice cap, which has also been observed, has the potential to produce coastal flooding worldwide.

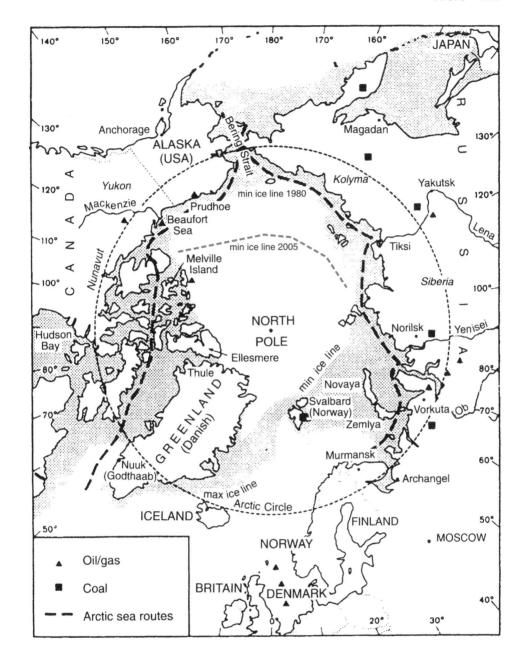

Tables

Table 1 Population

	Millions	% of world total
World	6,500	
China	1,300	20%
India	1,100	17%
EU 25	460	7%
United States	300	5%
Indonesia	250	4%
Brazil	190	3%
Pakistan	170	3%
Bangladesh	150	2%
Russia	140	2%
Japan	130	2%
Nigeria	130	2%
Mexico	110	2%
Other countries	2,070	32%

Table 2 GDP

US$ at PPP exchange rates		
	Total (billions)	Per capita
World	61,000	9,500
United States	12,000	42,000
EU 25	12,000	28,000
Germany	*2,500*	*30,500*
Britain	*1,800*	*30,500*
France	*1,800*	*30,000*
Italy	*1,700*	*29,000*
Spain	*1,000*	*25,500*
China	8,900	7,000
Japan	4,000	31,500
India	3,600	3,500

Russia	1,600	11,000
Brazil	1,600	8,500
Canada	1,100	34,000
Mexico	1,100	10,000
South Korea	1,000	20,500
Other countries	14,000	

Table 3 Share of World Economy (GDP)

	PPP	*Market exchange rates*
United States	20.4%	29.0%
EU 25	20.1%	30.9%
China	14.6%	5.2%
Japan	6.6%	10.8%
India	5.9%	1.7%
Russia	2.6%	1.7%
Brazil	2.6%	1.4%
Canada	1.8%	2.4%
Mexico	1.8%	1.6%
South Korea	1.6%	1.9%
Other countries	22.1%	13.4%

Table 4 Proven Oil Reserves

	Billions of barrels
World	1,300
OPEC 11	870
Saudi Arabia	*260*
Iran	*120*
Iraq	*110*
Kuwait	*99*
UAE	*90*
Venezuela	*71*
Libya	*37*
Nigeria	*35*
Qatar	*18*
Canada*	180
Russia	67
Kazakhstan	25
United States	24
Mexico	19
China	17
Other countries	98

* 90% oil sands, 10% conventional reserves

Table 5 Annual Military Spending

	Billions of US$	% of world total
World	1,100	
United States	520	47%
EU 25	210	19%
France	*45*	*4%*
Britain	*43*	*4%*
Germany	*35*	*3%*
Italy	*28*	*3%*
Spain	*10*	*1%*
China	81	7%
Japan	44	4%
Arab League 21	39	4%
Saudi Arabia	*18*	*2%*
South Korea	21	2%
Russia	20	2%
India	19	2%
Australia	18	2%
Turkey	12	1%
Brazil	10	1%
Other countries	110	10%

Related Information Sources

1 Periodicals and Government Sources

BBC Timelines. Online: news.bbc.co.uk/1/hi/world/ (summary of events for most countries and territories).

The Economist. Online: www.economist.com (world news and economic trends).

Europa World Yearbook. Routledge.

Jerusalem Report (Middle East events).

Statesman's Year Book. Macmillan.

UK Foreign and Commonwealth Office Country Profiles. Online: www.fco.gov.uk (select countries & regions).

UN Demographic Yearbook. Online: unstats.un.org/unsd/demographic

UN Peacekeeping. Online: http://www.un.org/Depts/dpko/dpko/ (description of current and past operations).

US Central Intelligence Agency *World Factbook*. Online: www.cia.gov/cia/publications/factbook (data on every country and territory).

US Department of State Background Notes. Online: www.state.gov/r/pa/ei/bgn/ (current situation in each country).

US Library of Congress Country Studies. Online: rs6.loc.gov/frd/cs/ (Books covering history, politics, demographics and economy for each country).

2 Other Atlases (latest editions)

Allen, John. *Student Atlas of World Politics*. McGraw-Hill, 2005.

Chaliand, Gerard, and Jean-Pierre Rageau. *The Penguin Atlas of Diasporas*. Penguin, 1997. (Demographics and migration.)

Chaliand, Gerard, and Jean-Pierre Rageau. *A Strategic Atlas: Comparative Geopolitics of the World's Powers*. HarperCollins, 1993.

Donald, Stephanie, and Robert Benewick. *The State of China Atlas*. University of California Press, 2005.

Duncan, Andrew, and Michel Opatowski. *Trouble Spots: The World Atlas of Strategic Information*. Sutton, 2000. (Focus on conflicts.)

Gilbert, Martin. *The Routledge Atlas of the Arab–Israeli Conflict*. Routledge, 2005.

Griffiths, Ieuan. *The Atlas of African Affairs*. Routledge, 1993.

McEvedy, Colin, and David Woodroffe. *The New Penguin Atlas of Recent History: Europe Since 1815*. Penguin, 2003.

Magocsi, Paul. *Historical Atlas of Central Europe.* University of Washington Press, 2002. (Detailed mapping of central and eastern Europe.)

Smith, Dan. *Penguin Atlas of War and Peace.* Penguin, 2003.

Smith, Dan. *State of the World Atlas.* Penguin, 2003.

Veregin, Howard (ed.). *Goode's World Atlas.* Rand McNally, 2005.

Index

Each number refers to a section (map and/or accompanying notes), not to a page. Former names are shown in *italic* in parentheses, e.g. Harare (*Salisbury*). Where appropriate, the number for the main entry is shown in **bold**.